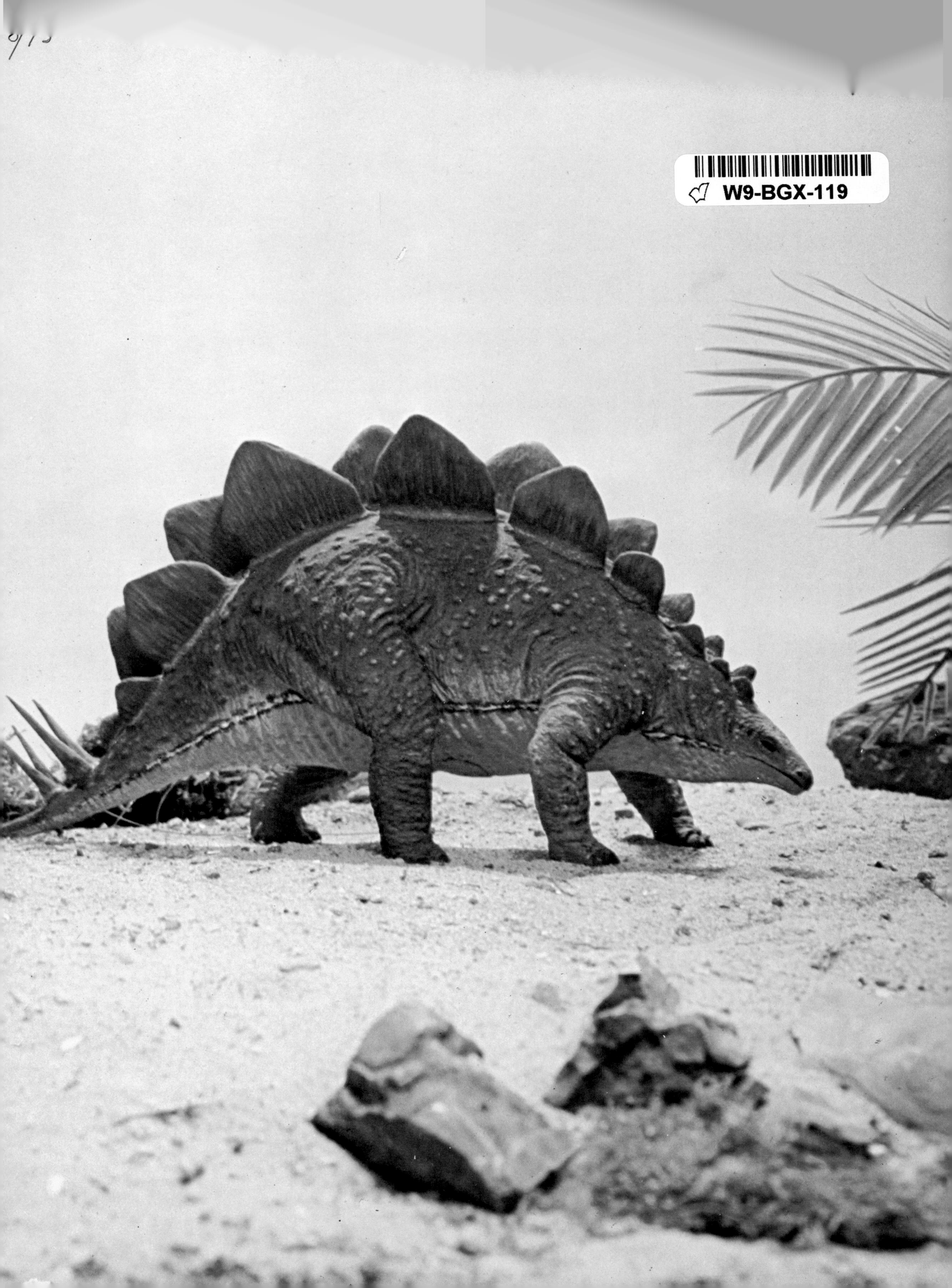

Prehistoric
Animals

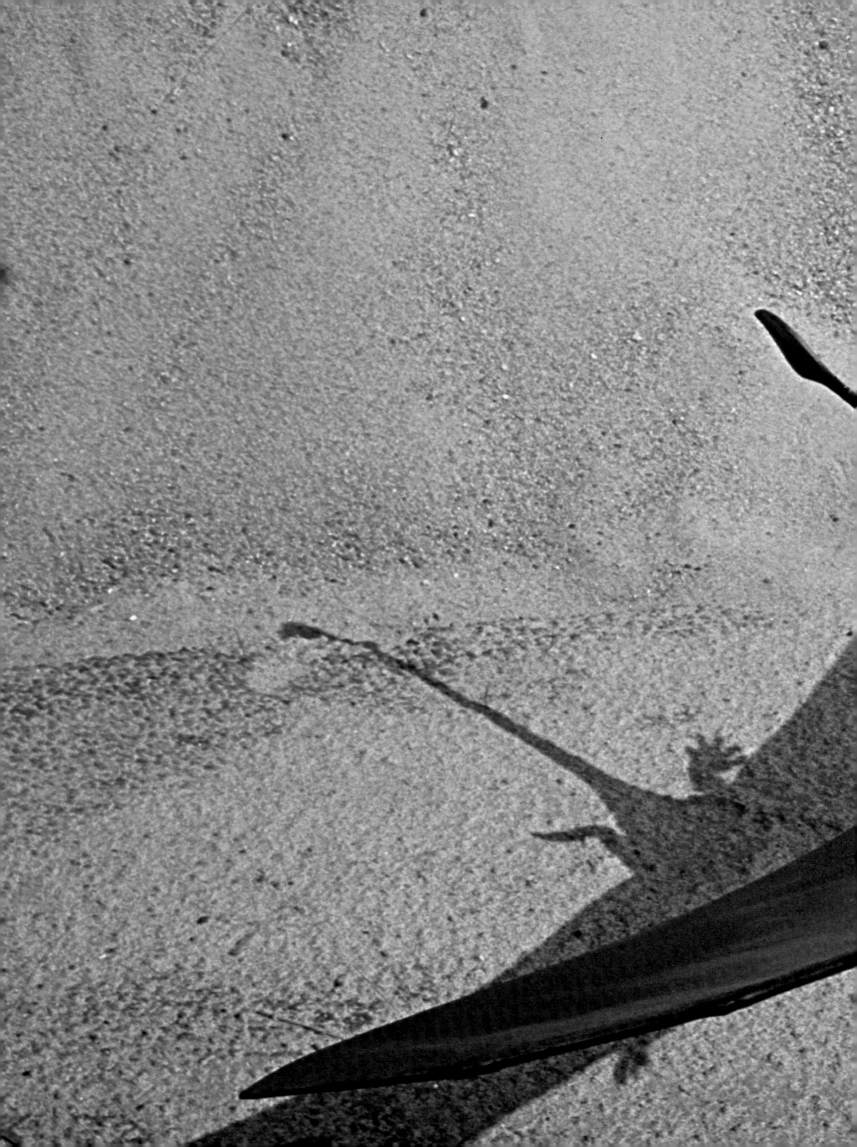

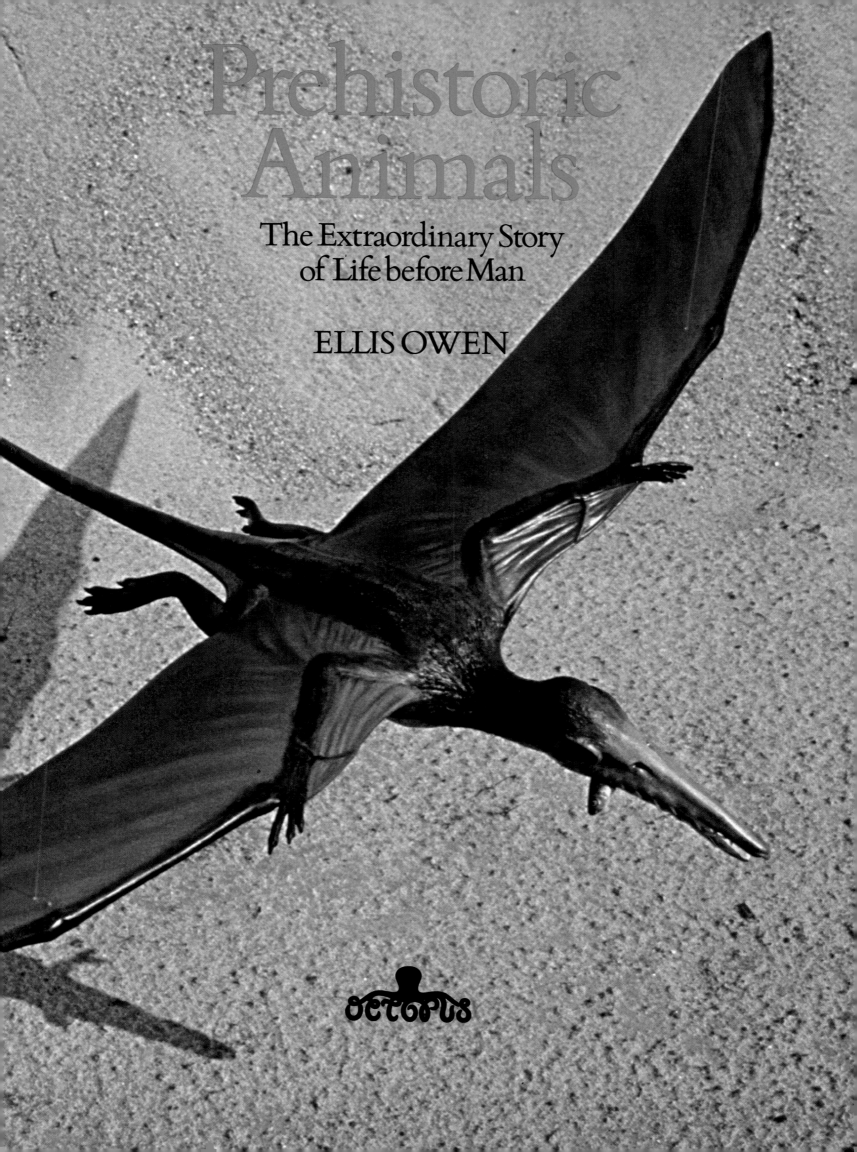

Prehistoric Animals

The Extraordinary Story of Life before Man

ELLIS OWEN

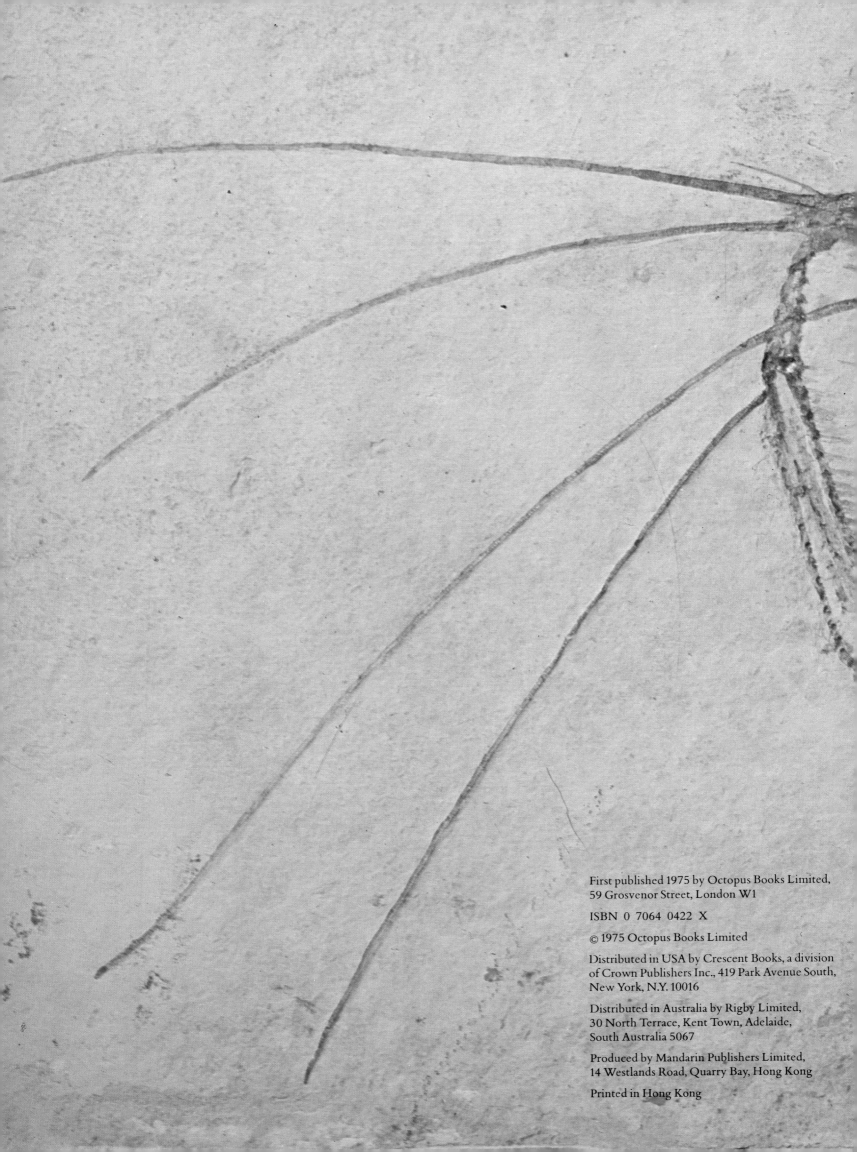

First published 1975 by Octopus Books Limited,
59 Grosvenor Street, London W1

ISBN 0 7064 0422 X

© 1975 Octopus Books Limited

Distributed in USA by Crescent Books, a division
of Crown Publishers Inc., 419 Park Avenue South,
New York, N.Y. 10016

Distributed in Australia by Rigby Limited,
30 North Terrace, Kent Town, Adelaide,
South Australia 5067

Produced by Mandarin Publishers Limited,
14 Westlands Road, Quarry Bay, Hong Kong

Printed in Hong Kong

Contents

Introduction

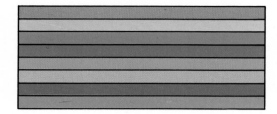

A quick look around the animal population of the world today might easily lead one to believe that the process of evolution had ceased and that all reptiles, mammals, birds and fishes had reached an optimum or ultimate stage in their evolutionary development. This is not the case, however, for evolution in both animal and plant kingdoms has been going on since the very dawn of life on earth. This continuous process is responsible for the ever-changing shapes of all forms of life, from the lowest single-celled animals to Man himself.

The rate at which each of the groups of animals or plants has evolved, or is evolving, varies considerably. While some groups have been in existence since earliest geological times and are still represented by living species today, others have evolved more quickly, have passed through many stages of adaptation or specialization in order to fit into a changing environment and have completed their evolutionary history. These groups form the bulk of the extinct genera and species which succumbed to the fierce competition for survival and existence.

They are not lost forever, for nature has provided her own history book bound by layers of sedimentary rock. Within these layers, or strata, the chapters of prehistoric life can once again be read. It is not in words that the story is told, but in the mineralized remains of the actual animals and plants themselves. They are the fossil skeletons and shells, the impressions, the footprints, the casts and moulds of a past era which we can find buried, often deep down, within the stratified layers of the earth's crust.

To study these remains more seriously, we must find out just a little more about geology, the science of the earth, and the other special subjects which scientists use. We must know, for instance, that the *Geological Column* (see opposite) represents a vast section through the various layers of the earth's surface, presented in chronological order. This column can be briefly classified into three main sections: the Paleozoic, dealing with early forms of life in the oldest sedimentary rocks; the Mesozoic, representing the middle layers; and the Cenozoic, which represents the youngest strata leading up to historic times.

Just how or why some animal groups became extinct can also be answered from evidence obtained from the layers of rock. The nature of past seas, their depth, and even their temperature and salinity, can all be explained by geologists with special knowledge of geochemistry, geophysics and sedimentology. From the results of their careful researches, these scientists can reconstruct the exact conditions existing in the world some hundreds of millions of years ago.

With such knowledge at our fingertips we can begin to understand the necessity for the numerous experimental shapes of many animals, particularly the somewhat grotesque body forms seen in the larger reptiles during the age of the dinosaurs. We can see how the horse has had to adapt itself to new surroundings of life on the plains instead of the woodland glades from which it originated. Perhaps we may also understand something more about the way in which some mammals have returned to the sea after having established themselves on land, and a little more about the gradual evolution of the birds from their original reptilian ancestors.

Before we can examine the evidence regarding the origin of life, we must first understand the ways in which the animal remains

have become preserved as fossils. There are many areas where sedimentary rock contains no fossils at all and yet we know that marine life must have existed there many years ago. What has happened to these remains? Why have they not withstood the many chemical changes which have taken place throughout time?
To find an answer to these questions, we must look more closely at the processes of sedimentation.

The gradual wearing down of land surfaces and areas of hard rock by the action of the wind and the rain is called erosion. The fine particles of rock, which are the product of erosion, are transported by rivers and streams to lakes and to the seas. Here they are deposited as fine sediments on the lake bottom or sea-bed, sometimes covering the remains of animals and plants and these eventually become changed into petrified fossils. The constant erosion of the land surfaces over many thousands of years may cause a marked change in the total weight in the land mass of a continent, with the result that, gradually, a continent may rise well above the level of the sea. Equally a sea-bed may sink because of the extra amount of sediment deposited upon it. As some land masses rise, others fall and are invaded by the surrounding seas. This process usually takes many thousands of years, unlike the more dramatic movements of earth masses by the action of earthquake and other subterranean disturbances.

Although most fossils that we find represent life in the seas and lakes, there are many exceptions. Large animals, such as the land-based dinosaurs and mammals, are often found in the position of death where their skeletal remains have been gradually covered by blown sands and dusts of the plains and deserts. Other creatures have been preserved in different ways. The asphalt tar pits at Rancho La Brea in California have trapped many an unfortunate grazing animal together with its carnivorous predator, presumably the latter having pursued its prey over the brink into the tar, and many well preserved skeletons have been removed from these pits during the subsequent workings by commercial firms.

The frozen wastelands of the northern Siberian Steppes also reveal their secrets from time to time. Several mammoths, which lived during the great glaciation of Pleistocene times, having fallen down crevasses or been covered by frozen rivers or lakes, remained in a state of 'deep-freeze', to emerge many thousands of years later in an almost perfect state of preservation. In one or two cases, both flesh and skin were so well preserved that dogs at the scene of excavation were able to eat them.

Insects have also become fossilized. Many a fine example exists of a bee, fly or beetle in amber. The amber, which in its natural state is sticky resin, entrapped an insect which remained engulfed until the resin changed, perhaps by the action of sea water or merely by the passage of time, to a hardened transparent lump. These are collected by some people in order to make beads, or other forms of jewellery.

For every fossil found in the strata, there were probably many more animals which did not become fossilized. The very nature of some animals did not allow their bodies to become changed or altered in any way and the soft tissues soon decomposed leaving nothing but a faint impression. In some cases not even an impression remains and the only reason we have for suspecting the existence of these creatures is that all the environmental and ecological conditions present then would have supported them.

The pearly *Nautilus* is closely related to the extinct species found in rocks from the earliest Paleozoic period to Tertiary times (see page 56). It is in fact a perfect example of a living fossil which has hardly changed at all since it first appeared. Some of these cephalopods, like their close relatives the extinct ammonites (see page 56), occurred as both coiled and uncoiled forms. The animal itself lived in the last chamber of the segmented shell shown here in the sectioned specimen of a *Nautilus pompilius*. A tube called a siphuncle connects each of the chambers which can be filled or emptied with gas by the operation of a siphon and thus the buoyancy of the animal is controlled.

	GEOLOGICAL PERIOD	CONDITIONS ON EARTH	ANIMAL LIFE
	The **CENOZOIC** Era 65,000,000 years ago	Renewed widespread volcanic activity; warm temperatures at beginning of era, then suddenly cooling to Ice Age conditions.	*Evolved:* Large running mammals, apes, man-like apes, man. *Flourished:* Reptiles, amphibians, fish, insects, corals, molluscs. *Died out (at end of era):* Larger running mammals, including woolly rhinoceros, woolly mammoth, sabre-tooth tiger, giant Irish deer.
	The **CRETACEOUS** Period 150,000,000 years ago	Little earth movement; temperatures gradually cooling towards the end of the period; marine swamping; first flowering plants.	*Evolved:* Gigantic dinosaurs, duck-billed dinosaurs, bipedal dinosaurs, marsupial mammals, placental mammals, snakes, beaked birds. *Flourished:* Small primitive mammals, reptiles, amphibians, fish, insects, corals, molluscs, echinoderms. *Died out (at end of era):* Many large reptiles, including dinosaurs, flying reptiles, fish-like reptiles, ammonites.
	The **JURASSIC** Period 200,000,000 years ago	Little earth movement; warm temperatures; humid, swampy conditions.	*Evolved:* Bird-hipped dinosaurs, herbivorous and carnivorous dinosaurs, armoured dinosaurs, first toothed birds. *Flourished:* Large reptiles, dinosaurs, small primitive mammals, ammonites, squid, echinoderms.
	The **TRIASSIC** Period 250,000,000 years ago	Little earth movement; sea flooding; desert conditions continue; alternating wet and dry seasons.	*Evolved:* Large marine reptiles, flying reptiles, fish-like reptiles, bipedal reptiles, ammonites, first dinosaurs, small primitive mammals. *Flourished:* Reptiles, amphibians, fish, insects, corals, molluscs, echinoderms, brachiopods.
	The **PERMIAN** Period 300,000,000 years ago	Little earth movement; very warm temperatures; desert conditions; widespread drying up of earth's surface.	*Evolved:* Herbivorous reptiles, more advanced carnivorous reptiles, land reptiles, freshwater reptiles, mammal-like reptiles. *Flourished:* Amphibians, fish, insects, corals, molluscs, echinoderms, brachiopods. *Died out:* Trilobites.
	The **CARBONIFEROUS** Period 350,000,000 years ago	Volcanic activity continues; tropical climate on much of earth's surface; large forests; swampy conditions.	*Evolved:* Primitive carnivorous reptiles, marine fish, first insects. *Flourished:* Amphibians, fish, sharks, corals, molluscs, echinoderms, brachiopods, trilobites.
	The **DEVONIAN** Period 420,000,000 years ago	Widespread volcanic activity; warm temperatures; arid desert conditions.	*Evolved:* Freshwater fish, shark-like fish, lung-fish, primitive sharks, first amphibians. *Flourished:* Corals, molluscs, echinoderms, brachiopods, trilobites. *Died out:* Graptolites.
	The **PALEOZOIC** Era 600,000,000 years ago	Early volcanic activity; cold conditions at beginning of era, gradually warming towards the end; no vegetation.	*Evolved:* Soft-bodied animals – jellyfish, worms, sponges; corals; shelled-animals – sea snails, molluscs; echinoderms – crinoids, starfish; sea scorpions; graptolites; brachiopods; trilobites; armoured jawless fish.

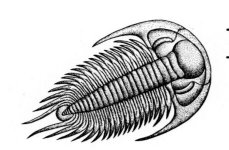

1 The Age of Trilobites

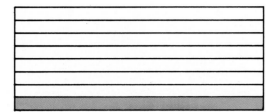

The Paleozoic Era

The tremendous length of time which has elapsed since the formation of the earth is difficult for the human mind to comprehend. Our concept of the early stages in the history of our planet is made even less comprehensible when we consider that, out of about 4,500 million years of the earth's existence, life, even in its simplest form, began only during the last 500 million years. It is not known for certain how carbon compounds, amino-acids and proteins came together to form living matter, but it took many thousands of years before these properties had reached such an advanced stage that even a faint impression could remain on the surface of the oldest sedimentary rocks. Many of these impressions were made by soft-bodied animals living in Pre-Cambrian seas, such as primitive worms, jelly-fish and very early segmented forms of life. It took millions of years before this situation showed any real sign of change.

By the beginning of the Cambrian period, other forms of life had evolved and the sea-bed was comparatively active with sea-snails or gastropods, bivalve molluscs, sponges, brachiopods and large segmented arthropods, called trilobites. There were also many of the primitive forms of life which had been predominant in the Pre-Cambrian seas, such as worms and jelly-fish.

The trilobites were very much the organized and active forms of life in these early seas. They had a body composed of three distinct sections or lobes: the cephalon or head, the thorax or body segments, and the pygidium or tail segments. The muscular soft parts of the animal were contained within a horny external skeleton or exoskeleton, rather like a beetle. Several pairs of jointed legs projected from the sides of the body, and the head-shield possessed a pair of antennae and, usually, although not always, two compound eyes with numerous lenses. The whole animal was well suited to its deep-water habitat and the muddy sediments of the sea-bed. Many species of trilobite evolved and they were represented throughout the Paleozoic rocks, only becoming extinct at the end of the Permian period.

Unlike the trilobites, which could move freely about the floor of the ocean, some animals attached themselves to objects, or remained firmly implanted on the sea-bed. These animals are referred to as *sessile*, or attached by their base. One of the most outstanding examples of this way of living is shown in the crinoid or sea-lily—a flower-like echinoderm related to the sea-urchins and starfish. It was formed from a cylindrical stalk or stem, upon which was surmounted a cup composed of flattened plates and surrounded by several long, often branching arms, which moved in wave-like motion with the currents. Several species of crinoid still exist today and underwater photography has enabled us to take a glimpse of what life on the sea-bed during early Paleozoic times must have been like.

Sessile organisms began to increase in numbers and species, including sponges and brachiopods, both of which have proved to be highly successful forms of life which still exist. Of these, the brachiopods are probably the least known and understood group. They comprise a strictly marine or sea-living group of very simple animals, possibly distantly related to the segmented worms. They are contained within two hard valves, sometimes consisting of two chitinous or horny plates, or they may be enclosed between two calcareous convex valves hinged together by tooth and socket

hinges. In this way they faintly resemble the molluscan bivalves although they are in no way related to this group.

A hundred million years had passed since the beginning of the Cambrian period and now the scene was changing. During the Ordovician period which followed, most of the early organisms which had already established themselves in Cambrian times began to develop even more quickly. Alongside the bivalve molluscs and brachiopods there appeared another molluscan form. This was altogether different in shape and habit. It looked like an octopus but lived in a long, straight conical shell. There were two common genera in Ordovician seas; one was *Endoceras* which is known to have reached 10 feet (3 m.) in length and resembled the present-day giant squid. The other genus was *Sactoceras*, slimmer, more lightly built animals which propelled themselves along at speed by using their water-jet system of propulsion. Their development at this time was an important step in the long trail of evolution of the Mollusca and we shall see that this group becomes an even more important marker in Mesozoic times, for the zonation, dating and classification of the strata.

Other molluscs also increased in numbers and diversity. Gastropods and bivalves produced species after species at an alarming rate, and corals and brachiopods were also in abundance by the end of Ordovician times.

This dramatic burst of evolution among the simpler animals continued in the seas right through to the Silurian period, some seventy-five million years later. During this time some species became extinct while others continued to develop and adapt themselves to gradually changing conditions of increased temperature and differences in salinity.

Conditions certainly favoured the brachiopods. In the shallower waters near shore lines and beaches where the calcium concentration was at its highest, these animals increased to such an extent that often banks of brachiopod shells were formed, in very much the same way as happens in bivalve molluscs. In fact, the Silurian was a time in which the brachiopods were at their most abundant. Many genera and species were produced and have proved very useful horizon markers to geolgists dealing with the classification and naming of zones and strata.

Almost twenty million years had gone by since the beginning of the Silurian period and many changes had taken place. The crinoids had increased in numbers and species and jelly-fish were still pulsating their way around the murky depths. Hundreds of trilobites had developed and there were many new arthropods, such as the eurypterids or large sea-scorpions. But an even greater and more dramatic development had occurred. The first fish had evolved, and with it a major step in evolution had been achieved. Instead of the external skeleton of the trilobites and other arthropods, animals had now developed an internal framework which supported their muscles and protected their vital organs. This framework was held together by a central column called a vertebral column; and so fish became the first true vertebrates.

The earliest forms were the jawless fish, or agnathans, very simple tube-like animals with dorsal and lateral fin-flaps and simple openings at the mouth and the anus. An example of this very primitive form is *Jamoytius* which lived in European and possibly Scandinavian Silurian seas.

Previous page By the middle of the Ordovician period the pattern of trilobite morphology had become very variable and some extremely fascinating and delicate shapes appeared among the many diverse species. *Ceraurus*, shown here as a reconstructed model, was a large species with lateral extensions of its head-shield, well-developed antennae and a pair of compound eyes. It was approximately 6 cm. (2⅜ in.) in length.

Below In early Paleozoic lakes and streams the very earliest forms of fish life began their evolutionary history. But these were not the typical fishes which we can recognize in the succeeding periods. Instead they resembled the lampreys of our modern depths. These primitive forms were the ostracoderms and were represented by the simple *Jamoytius*, *Cephalaspis* and *Pteraspis* as well as many other genera. The typical fishes were represented in these early lakes. *Climatius* and *Gemuendina* are among the more readily recognizable.

Right Although most species of trilobite which evolved throughout the Paleozoic Era were not particularly large animals, *Paradoxides*, which lived during the Cambrian period, was an exception. It reached a length of approximately 50 cm. (19⅝ in.) and had an enormous cephalon or head-shield, and large glabella. Some of the finest examples have been found in Middle Cambrian rocks of Massachusetts, U.S.

Right Trilobites are exclusively Paleozoic in age, ranging from the Lower Cambrian to Permian period and reaching a peak in their evolution towards the end of the Ordovician. The evolutionary changes throughout their history are illustrated in the very distinctive patterns of morphology or general outlines. It is possible for some paleontologists to determine, merely by glancing at a trilobite specimen, from what period it originated. The specimen shown here is *Ellipsocephalus hoffi* and comes from the Cambrian of Czechoslovakia. It is 3·3 cm. (1¼ in.) long.

Opposite The reconstructed sea-bed scene from the Lower Cambrian shows vertical calcareous tubes containing primitive marine worms which displayed tentacle-like fans or arms which they waved about in order to catch particles of food in the surrounding water. In the foreground is the shell of a very primitive limpet-like mollusc *Scenella* which had a bilateral symmetry.

The conditions *Top left* in which certain groups of animals live often has a great influence, not only upon their behaviour, but also upon their shape or form. Trilobites provide a perfect example of a group which has been greatly influenced by the ecology of their environment. The species seen here is *Lloydolithus ornatus* from the Middle Ordovician of Britain and shows the proportionately large head and low-lying cephalic margin with elongated lateral spines. These features allowed the animal to disturb vast quantities of muddy sediment on the sea floor by grubbing around or shuffling with the minimum of effort. It was from these disturbed sediments that it drew its food. It was approximately 15–20 mm. (6–8 in.) long.

Top right *Dinorthis porcata* from the Middle Ordovician of Anticosti Island, Canada. The three specimens grouped here are typical examples of articulate brachiopods from the Lower Paleozoic. The strong radiating ribs or costae originate from the umbo (apex) situated midway along the hinge-line. This character alone prevents the brachiopod being confused with bivalve molluscs which do not have the umbo situated in a central position along the hinge, but often placed with a bias towards one side or the other. The individual specimens are approximately 4 cm. (1½ in.) wide.

Right Graptolites were organized filamentous colonies of zooids which produced an abundance of buds and branches. Some of these organisms remained stationary, almost plant-like objects implanted in the sea-bed, while others, such as the *Didymograptus murchisoni* species shown here, attached themselves to floating weed or drifting objects and remained suspended in the water. This species lived in the Ordovician period and can be found, though rarely, in the shales of that age in Shropshire, England, and parts of North Wales.

Left It is not surprising, when we look at some corals, that they were once regarded as plants. The method of growth of some genera, with long branching side shoots and cup-like heads, might well have led some observers to this conclusion. Some forms even have root-like structures at their base in order to help them remain upright. In fact, corals are simple animals related to sea anemones and are classified with the coelenterates. They can be separated into different subclasses according to their method of living and their morphological features. The Silurian coral shown here belongs to a group of simple corals which lived apart from reefs. It is about 8 cm. (3⅛ in.) long.

Below left There was not much diversity of shape within the range of Lower Paleozoic corals. Most species were either conical or long branching varieties. The *Omphyma turbinata* seen here is typical of the low conical coral from the Wenlock Limestone of the Silurian rocks of Dudley in Worcestershire. It is 8 cm. (3⅛ in.) in diameter.

Below Trilobites were rather special arthropods, or segmented animals, which lived in the murky depths of Paleozoic seas. It is probable that they preferred the softer and finer sediments on the sea-beds of the coastal areas. They are called trilobites because they have segmented bodies which can be clearly divided into three lobes—a central column, or glabella, separating two lateral lobes. The soft parts of the body were enclosed within a tough horny or chitinous exoskeleton consisting of a semicircular head-shield or cephalon, a central segmented thorax and a grooved pygidium or tail. Some species were without sight, but others had rudimentary compound eyes rather like those of a fly. Two whip-like antennae were also highly sensitive sense organs. The specimen shown here is called *Ogygiocarella debuchi*, and it was quite frequently found in the sea during the Silurian period. It was approximately 5 cm. (2 in.) in length.

Bottom right *Jamoytius* was a very primitive fish with a long slender streamlined body which had lateral and dorsal fin-flaps for stability. It also had quite well developed lateral eyes and a terminal mouth. Not very many well preserved examples of this tube-like fish have been found, and a great deal of the published descriptions have been based on impressions and fragments. It probably swam quite swiftly and may have attained a length of about 25 cm. (10 in.). The specimen figured here is a reconstructed model, although several specimens have been collected from Silurian rocks.

Right Trilobites were related to the earliest segmented worms but, by the middle of lower Paleozoic times, had become quite organized, and many fine species had evolved. The specimen shown here is of *Proeteus*, a comparatively common genus from the Wenlock Limestone of Dudley, Worcestershire, England. This neat-looking example is about 2 cm. (¾ in.) long.

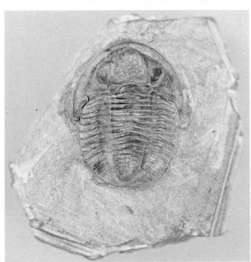

Left The usual idea of a sponge is the soft object which we use in the bathtub. Sponges are, however, real animals which have been classified as Porifera, which means 'pore-bearers'. The skeletons of these animals consist of a network of either calcareous or siliceous spicules which support the animal. When the animal part of the sponge dies, all that is left is the skeleton. Fossil sponges are not always easy to recognize and the amateur collector may be forgiven if he should mistake the figured specimen for just another amorphous nodule. It is in fact a good example of *Astylospongia praemorsa* represented in the Middle Silurian rocks of the island of Gotland and the U.S. When cut in half and polished, the silicified inside looks astonishingly beautiful. Most specimens are about 4 cm. ($1\frac{1}{2}$ in.) in diameter.

Right Early Paleozoic seas had their own characteristic fauna of trilobites and brachiopods, molluscs and other invertebrates. Among the fauna were many beautiful sea-lilies or crinoids which remained almost unchanged in general shape throughout their evolutionary history. The specimen shown here is a typical Silurian species, *Dimerocrinites speciosus* from the Wenlock Limestone of Dudley in Worcestershire. It is approximately 14 cm. ($5\frac{1}{2}$ in.) long.

Above The semicircular head–shield and flattened body, shown here in the specimen of *Cephalaspis lyeli*, were obviously developed as a result of its environment. The lengthy tail and lack of lateral fins added greatly towards keeping the fish on the muddy bottom of the lake. It became more widespread in Devonian times ranging from Europe to Asia and North America.

Left It is not surprising that *Pterygotus* has been called a sea–scorpion, but there the analogy ends, since this particularly cruel-looking Eurypterid was between 2–2.5 m. (6½–8 ft.) in length. It carried a pair of wicked looking toothed pincers in front of its huge flattened, segmented body. Living mainly upon the sea floor in late Silurian and early Devonian times, this vast arthropod, which could swim for long distances, was a dangerous and deadly hunter of the deep.

Opposite *Cephalaspis* originated in the Silurian period but was further developed in the Devonian freshwater lakes and streams where it spent most of its time grubbing about among the softer sediments on the bottom for food particles. It extracted these by sifting mouthfuls of mud.

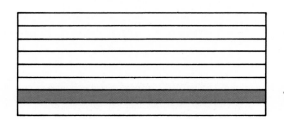

2 The Age of Fishes

The Devonian Period

This is known as the age of fishes because it was during this period that tremendous changes took place among the primitive jawless fish, the Agnatha, and other early fishes. Some of these early forms developed large bony plates around their head regions and were known as Ostracoderms. Two of the best known examples are *Cephalaspis* and *Pteraspis*, which were fairly common inhabitants of the streams and freshwater lakes in Lower Devonian times. Although fossil remains of these genera are found in the Old Red Sandstone strata of equivalent age in Europe, Canada and Australia, early examples of the group have also been found in Silurian rocks of South Dakota, Colorado and Wyoming in the U.S. Other genera developed at about this age were *Anglaspis*, a smooth oval form, and *Birkenia* which had a row of hook-like, torpedo-shaped spines on the back and long pointed spines on the under surface, as well as the thick bony head-shield for protection.

This heavily armoured type of fish existed for some time, but the process of evolution produced a series of changes in its basic anatomy. The next stage saw less plating and more mobility of the head. A peg-like articulation of the head with the rest of the body now gave rise to another group, the arthrodires, represented by *Coccosteus*, a small genus with bony plates inside the mouth to serve as simple teeth. Later, *Dinichthys*, which varied in length from 3 to 30 feet (0·9–9 m.), developed sharp pointed teeth instead of the simple bony plates. Although this was a strong, fierce group, it did not survive and soon became extinct.

Specialization of form continued as each group reacted to the changes in the environment and ecology. Some adapted while others failed. Among the more highly specialized members of the armoured placoderms was *Bothriolepis*, which occurred during Middle and Upper Devonian times. It is often referred to as the 'box-fish' because its large rhomboidal body plates were arranged in an almost box-like form. It had long lateral arm-like appendages, which assisted in stabilizing it when swimming, and a well developed dorsal fin and a long tail.

Towards the end of Devonian times many fish moved down the streams and rivers and adapted themselves to semi-marine and marine conditions. One group included genera which developed some of the characters of present-day sharks and although they were completely unrelated to this group, they established themselves as fierce predators of smaller forms of marine life in very much the same way as the sharks. They had numerous spines, and spiny fins and for this reason were classified as acanthodians. The genus *Climatius*, with species only 3 or 4 inches (7·6–10·2 cm.) long, is a typical example of this group.

The true sharks were represented in Devonian times and appeared towards the end of the Upper Devonian period. A good example is the genus *Cladoselache* which attained a length of 4 feet (1·2 m.), and had numerous sharp triangular teeth. Some of the sharks and rays which live in our seas today are directly developed from these early forms. Their success was largely due to their ferocious nature and immense strength.

With the general rise in temperature towards the middle of the Devonian period, lakes and rivers began to dry up. Such conditions produced an added problem of survival for some of the freshwater fish faunas. Many species succumbed to the rapid change in conditions and died out altogether. Others began to adapt

themselves by using their fleshy fin bases as a means of locomotion from one muddy pool to another. In other words, fish were beginning to walk. The 'fringe-fins', as they are generally known, were the earliest example of a group which eventually gave rise to the amphibians. A typical member of this group is *Osteolepis*, a small slender fish, no longer than 7 to 8 inches (17·8–20·3 cm.) and with a short, blunt skull composed of several pairs of rhomboidal-shaped bones.

A well known group closely related to the 'fringe-fins' is the sub-group of Coelacanths, which further developed in Jurassic seas and, until comparatively recently, was thought to have become extinct in the late Cretaceous period. However, in 1938, a living specimen was caught by fishermen off the South African coast. The existence of an 'extinct' fish gave rise to much discussion and speculation regarding the possible existence of other fossil animals.

In addition to the walking fish there were also those which adapted themselves to breathing air through a very simple lung, as well as absorbing oxygen through their gills. They were distantly related to the 'fringe-fins' and became known as 'lung-fish'. They could survive out of water by using the swim-bladder, or stabilizing organ, as a sack containing air in which an exchange of gases could take place, in much the same way as a normal lung. At the same time they continued to use their gills when in water and so could survive good and bad conditions. A well known example of this group is *Dipterus*, which was fairly widespread in Lower Devonian times.

As a result of adaptation to such widely differing environmental changes throughout several thousand years, a transformation had occurred which was the beginning of a huge trend to life on the land. Now the fish had developed the power to breathe the air of the atmosphere, and to use crude or rudimentary limbs for terrestrial locomotion. The next stage was to encompass both these powers in a partial existence on land and in water. The amphibians in fact achieved this stage and developed it to perfection. One of the earliest examples of this fascinating group was *Ichthyostega*, a sluggish, probably scaly skinned salamander-like animal approximately 4 or 5 feet (1·2–1·5 m.) in length, of late Devonian times. Examples have been found in deposits of Upper Devonian shale in East Greenland. We shall follow the further development of this strange group during the section on the Carboniferous period (see page 28).

Throughout the Devonian period, the development and evolution of the fishes progressed at a considerable rate, while the less spectacular invertebrate faunas continued to thrive and evolve at a more uniform pace. The large trilobites of the earlier Paleozoic Era were now being replaced by more diverse forms. Such genera as *Dalmanites* and *Calymene*, which probably began in the late Silurian seas, were still in existence and other genera were already fast evolving. Many of the species which had thrived in the Silurian period, however, had become extinct.

Likewise, the great burst of brachiopod genera and species seen in the Silurian period was declining slightly, but new forms, which had developed towards the end of that time, were becoming established. *Mucrospirifer* and *Atrypa*, for example, began in the Silurian or developed from Silurian species and had now established themselves in the Devonian. These delicately ornate brachiopods

must have been quite a contrast to the dreary forms of undersea life which were in existence at that time.

This was also the age of the corals and many genera and species were evolved. Some of these were reef formers, while others led a separate single existence, seemingly free of colonial protection. We find such interesting developments as cup-shaped corals with lids such as *Calceola sandalina*. A conical variety, *Streptelasma*, remained almost unchanged from Ordovician to Devonian times. A very successful type of coral, it appears, almost identical in shape, in Carboniferous and some Mesozoic rocks.

Crinoids, or sea-lilies, were also in abundance and many fine species flourished. Some had beautiful flower-like heads consisting of numerous branching arms and must have looked particularly graceful in their underwater garden. These animals are related to sea-urchins and starfish which also abounded in the Devonian seas.

In the freshwater inland lakes and rivers, molluscs had developed and large mussels or clams were fast evolving. Surrounding the lakes and becoming more widespread were the numerous subtropical plants which were to progress further and more rapidly in Carboniferous times.

Above Corals can serve as very useful index fossils. One of these is *Calceola sandalina*, a rugose coral which occurs in the Middle Devonian. It is a very distinctive species, maintaining a regular triangular shape and having a hinged lid or operculum which protected the soft parts of the animal at the calyx. The species occurs in Europe, Asia, Australia and North America. The average width of this form is approximately 5 cm. (2 in.).

Opposite The extraordinary adaptation of the 'lungfishes' to adverse conditions was surprising enough and, as the necessity to move from one disagreeable environment to a more pleasant or safer one became more urgent as conditions became worse, new methods of adaptation were tried. The crossopterygian fishes managed to adapt themselves to such adverse conditions by developing more muscular fin-bases which could be used as primitive limbs aiding terrestrial locomotion. It was in this way that a trend towards life on land as well as in the water began during Devonian times. *Eusthenopteron* was an early ancestor of the amphibians which were to master this way of living. It was about 60 cm. (2 ft.) long. long.

Right Not all of the crinoids were stout-stemmed robust animals. Some, such as the *Parisocrinus zeaeformis* figured here from the Lower Devonian slates of Bundenbach, Germany, were fine filamentous forms with an extremely thin stem and delicate, many branched arms surrounding an elongated body. This remarkable assemblage shows the specimens as they occurred in life. Each of the largest individuals seen on the slate slab is approximately 5 cm. (2 in.) long.

Below right One member of the arthrodire fishes was the flattened *Gemuendina* which is found in the Lower Devonian sandstones of parts of Germany. It is a long-ranging genus, being represented by many species throughout the Devonian, not only in Germany, but also in North America. As is the case with most flat fish of the time, it was equipped to remain fairly inactive on the bottom of lakes where it must have been almost invisible to its predators.

Previous page A very fierce member of the class Arthrodira, fish which had jointed necks, *Coccosteus* lived in the freshwater lakes of the Upper Devonian in Europe and North America. It was one of the largest armoured species, having a broad blunt head with rows of short but extremely sharp teeth in its wide mouth. A larger member of the same class was *Dinichthys* which is reputed to have reached about 10 m. (32¾ ft) in length.

Left The group of short horn-like corals shown here is typical of the many species of rugose coral found in the Devonian. These specimens come from Ontario, Canada, and show the numerous septa, or walled divisions, comprising the calyx in the shallow cup at the top of the coral. The larger specimens are 8 cm. ($3\frac{1}{4}$ in.). long and 5 cm. (2 in.) in diameter at the calyx. Each small cone was part of a large branching tree-like colony.

Below As yet there is no actual evidence of the steps in the transition from fins to legs, but the earliest example of a primitive amphibian is from the Upper Devonian of East Greenland where several skulls and other parts of the skeleton of the earliest labyrinthodont *Ichthyostega* have been found. The skulls are most important because, from these, we are able to learn more about the affinities which exist between this amphibian and some of the fishes which had shown evolutionary trends towards the amphibians' more advanced form of life. The living animal would have had a well developed long caudal fin.

Opposite top left There is a marked difference between the sponges of the Paleozoic and the Mesozoic Eras. *Hydnoceras tuberosum* is an extraordinary looking animal, but typical of the bizarre shapes displayed by some groups of Devonian sponges. It consisted largely of a tubular mesh of spicules which were the skeletal support for the delicate animal. The species is by no means common but is sometimes found in the Upper Devonian limestones of western New York State, U.S. The specimen shown here is 20 cm. ($7\frac{7}{8}$ in.) long, but some are over 30 cm. ($11\frac{4}{5}$ in.).

Opposite top right There was quite some variation of general shape among the ostracoderm fishes and some of the varieties were very successful. One of these, a member of the Anaspid group, was *Birkenia*, a small but somewhat bizarre form which had a row of hook-like dorsal spines and a tough, coarse scaly skin. Unlike earlier ostracoderms, it was better equipped for swimming and probably spent much of its time in the deeper water rather than grubbing around on the muddy bottom of a lake or stream. The model is about 6 cm. ($2\frac{3}{8}$ in.) long.

Below right Among the armoured jawless fish of the Devonian period was the genus *Pteraspis* which in some ways was a similar type of fish to *Cephalaspis*, but had a more streamlined body with a long tail-fin and sharp pointed dorsal spine. Its snout was elongated or pointed and it had deep-set lateral eyes. Most of the posterior part of its body was covered in scaly skin. The general shape of the body and the elevation of its tail-fin suggest that it spent some time in deeper water and also dwelt very largely upon the muddy bottoms of lakes and streams. *Pteraspis* has been found in the Old Red Sandstone of the Devonian rocks of England, and at an equivalent horizon in Germany, parts of Scandinavia and the U.S. The model is about 25 cm. (10 in.) long.

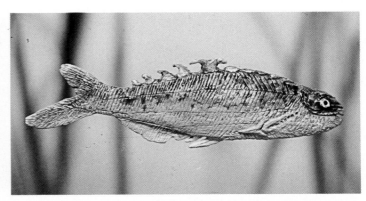

Previous pages Towards the later stages of the Paleozoic Era, the fish genera and species became very much more diverse with such forms as *Bothriolepis* and *Dinichthys* among the heavily armoured species with the typical fish represented by the lungfish, *Dipterus* and fringe-finned fish, *Eusthenopteron*. The sharks were represented by such genera as *Cladoselache* which reached a length of 2 m. (6½ ft). The fish genera which had begun their evolutionary history in the murky depths of large muddy lakes were now beginning to move down the streams towards the seas, where they would have to adapt themselves to completely new conditions.

Left Of the numerous species of trilobite evolved during the Paleozoic Era, few have been as well preserved as those found in Devonian rocks. The specimen seen here is of a tail or pygidium of *Odontophile hausmanni* from the Lower Devonian of Czechoslovakia. It shows the very clearly marked grooves and the base of the glabella.

Left Some trilobites had the ability to fold up in very much the same way as the modern woodlouse. This was obviously a very effective means of protection from predators. It is possible also that they assumed this position when conditions within their own environment became unfavourable. The specimen shown here is *Phacops milleri* from the Upper Devonian, Silica Shales, Erie Co., Ohio, U.S. It is about 4 cm. (1½ in.) wide.

Below left Echinoderms have been in existence since the earliest Paleozoic Era and have changed very little in morphology and function throughout their long history. The specimen seen here is *Palasterina follmanni* from the Lower Devonian slates of Bundenbach, Germany, and shows the characteristic ornament upon the upper body surface of this animal. The specimen is about 8 cm. (3⅛ in.) wide.

Opposite above Another fish which dwelt largely on the bed of freshwater lakes is *Bothriolepis*, a magnificent box-like animal enclosed within strong symmetrical plates and having a long tapering tail. Two comparatively long lateral arms or appendages must have given the fish great stability. In the same way as *Cephalaspis*, it grubbed around on the muddy bottom, sifting through the sediments in search of food. The model shown here, however, seems to be enjoying a somewhat more sandy environment.

Opposite below The very changeable climate which prevailed during the middle to late Devonian times forced many fish to adapt to a different environment. There was a tendency for lakes to dry up suddenly or, just as easily, flood. Such instability was dealt with by some species quite simply by developing a method for breathing air as well as absorbing oxygen through gills. A rudimentary sack-like lung was adapted from a stabilizing bladder. This enabled an exchange of gases to take place during extreme drought. In this way respiration was maintained both in and out of the water. *Dipterus* was one of the earliest 'lungfishes' which survived in this way. The model is shown here in its rather more familiar aquatic habitat. The species probably reached 2 cm. (¾ in.) in length.

3 The Birth of Amphibians

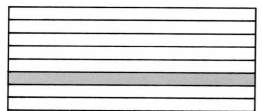

The Carboniferous Period

Most people will be familiar with at least one fossil from the Carboniferous period which gets its name from the numerous deposits of carbonized plant remains or coal. It was not until the end of this particular stage, however, that the steamy swamps and dense forests became widespread and prolific; and these forests have proved a major asset to the developing industrial world. The richest coal deposits are to be found in the U.S., Britain, the European continent and Australia. It is in these continents that we find evidence of the animals which lived within the swamps and forests, for it was here that terrestrial life was born. Behind this curtain of time a new experiment was about to begin.

The shallow warm seas which surrounded most coastal areas of the world in Upper Devonian times continued throughout the Lower Carboniferous period and gave rise to an abundance of invertebrate animals. Countless new genera and species of bryozoa, crinoids, corals, molluscs and brachiopods flourished as they had done in Silurian and Devonian times. Small colonies of non-reef-forming corals assumed remarkable shapes. Some were like elongated cones or horns with a head or calyx consisting of low radiating plates, arranged around a central column to form a shallow cup. These were called rugose corals and included many different types and species. A good example of a solitary form is *Zaphrentites*, a fairly common genus in the deeper waters of Lower Carboniferous seas. Other examples include *Lonsdaleia* and *Lithostrotion* which probably lived in the warm shallow waters nearer the coast.

Reef corals were also very common during this period and many limestones deposited at that time contain all the faunal assemblages of reef conditions. Typical of these were many of the smaller brachiopods with smooth shell surfaces, such as *Dielasma* and *Composita*.

Some brachiopods lived in fairly deep water on muddy sea-floors and required either long spines or extended hinge-lines in order to prevent themselves from sinking into the soft mud. The spiny species nearly all belonged to the group known as the Productoidea, the largest representative genus being *Gigantoproductus* which was a very large concavo-convex shell with an extensive hinge-line sometimes 12 to 14 inches (30·5–35·6 cm.) in width. Though they can never be regarded as common fossils some species are found in Lower Carboniferous limestone rocks of England and Wales. Associated with these giant brachiopods are often smaller, more delicate and, in many cases, more highly ornate species. One of these, *Spirifer striatus*, had a wide shell with numerous fine ribs or costae upon the surface which radiated out in fan-like fashion. Another interesting genus found in the Carboniferous Limestone is *Pugnax*. Many other brachiopod species occurred during this particular period which eventually gave rise to genera and species in the Permian seas. We shall see what remarkable shapes they assumed and the extraordinary way in which some forms imitated the corals.

This was also a time of quiet conditions and warmer waters and most forms of life on the sea-bed seemed to be increasing in size. The abundance of lime in the water added greatly to the variability of the species and produced optimum conditions for such forms as the crinoids or sea-lilies, which grew to enormous proportions with vast heads crowned with waving arms. Some very fine examples of this curious animal are preserved in the crinoidal limestones of North America and Europe and include the genera *Cyathocrinites*, *Taxocrinus*, *Woodocrinus* and *Platycrinites*.

By about the middle of the Lower Carboniferous period, the fishes had established themselves as dominant marine animals and were well adapted to their environment. Primitive sharks, such as *Xenacanthus*, marauded the seas off America, Europe and Australia and were to prove themselves a highly successful group. Although the skeletons of these fish were composed of cartilage which never became fossilized, we have been able to gather a substantial amount of information regarding their shape and general characters from the impressions and markings of fins and spines and many teeth, which is practically all that remains of this group. From this data, it is safe to assume that the overall shape of the shark, as we see it today, has changed very little throughout its long history.

The fish which we had seen developing rather specialized features in the Devonian period had, by the middle of Lower Carboniferous times, advanced in their evolution and had given rise to another special group called Rhipidistians. These differed from other related groups in having fewer bones comprising their skulls and which were arranged in such a way that internal and external rudimentary nostrils were developed. This was another step towards adaptation for air breathing. In fact, what had begun in the Devonian was now being completed in the Carboniferous. Amphibians, evolving from such genera as *Ichthyostega* in the Devonian period, had literally established a foothold and were evolving along characteristic lines.

At the end of Carboniferous times, the swamps and glades became the ideal conditions for the establishment of amphibian life. Within the surrounding forests hordes of insects provided one source of food, while in the swamp waters themselves masses of fish, molluscs and other aquatic animals abounded.

It was not surprising that, under the uniform climatic conditions of the Upper Carboniferous, the amphibians began to thrive and spread to all parts of the world. Many new genera and species appeared; some were more like alligators or crocodiles, with massive lumbering bodies and thick bony skulls. Others, such as *Megalocephalus*, developed large and odd shaped heads and, probably as a result of such experimental shape, eventually died out. While some species were rapidly evolving, others remained comparatively unchanged, apparently content with their environment. Large salamanders, representing the true amphibians, such as *Eogyrinus*, which lived in Upper Carboniferous swamps, showed all the characteristics of a mainly aquatic existence. It had weak limbs and a deep, smooth powerful body and long tail.

The development of the reptiles from their amphibian ancestors is still not completely understood. Somewhere along the line of amphibian evolution there is a link which would connect this line to the reptiles but this link has not been firmly established. Many paleontologists over the years have expressed both approval and doubt about the theories advanced regarding the origin of the reptiles, and one opinion, which is still held by some authorities, is that the Seymouriamorpha, represented by the genus *Seymouria*, provides such a link. However, recent developments in research into the origin of the reptiles does not support such a theory, although it offers no tenable alternative to substantiate its doubts.

Marked differences in the arrangement of skull bones and the type of teeth found in the jaws of some of the earliest reptiles have given the main clues as to the direction in which evolutionary trends lead but, as yet, there seems to be no clear pathway to follow.

Previous page The larger salamander-like amphibians, which began to appear in the late Carboniferous forest swamps, gave rise to new genera such as *Eryops*, a model reconstruction of which is seen here. It was a thick-bodied animal about 2 to 2½ m. (6½–8 ft.) in length, with a broad flattened head, and short legs which seemed inadequate to support its body. As a result it probably did not move far from the water's edge, where it spent its time lying in the sun. *Eryops* was by no means common and it is only very seldom found in Permian rocks of the U.S.

Above left Sea-lilies or crinoids, such as the *Paterocrinus multiplex* figured here, were still very much part of the sea-bed fauna in the Upper Carboniferous period. This beautiful calyx with thick branching arms is preserved in a pure white limestone from Mjatschkowa near Moscow, U.S.S.R. It is approximately 4 cm. (1½ in.) in width and 5 cm. (2 in.) high. The ossicles making up the stem of the crinoid are often of a distinctive pattern or shape and their presence in limestones will usually give the geologists some indication of the age of the rocks they are examining.

Below Thick forests of the Upper Carboniferous deltas and swamps provided a perfect setting for insects. Although few species had developed by the end of this period, one in particular always figures in Carboniferous forest scenes, *Meganeura*, an enormous winged insect similar to a modern

dragonfly. Its wingspan was just less than
1 m. (3¼ ft.) and the fast movement of the
wings of this giant must have created a
tremendous noise. It is not absolutely certain
whether or not these giant insects were
related to the true dragonflies.

Top Among the group of fish which are
known as the Dipnoans were several
advanced species with thickened fin-bases
which were really nothing more than
primitive limbs enabling the fish to move on
solid ground. *Uronemus*, a model of which
is shown here, also had a very crude lung in
addition to the gills, and was therefore able
to breathe both in and out of the water.
This type of fish was very well equipped to
withstand the most extreme conditions of
drought and could even move short
distances on land from a poor environment
to a better one. Most species were
approximately 10 to 12 cm. (4–4¾ in.) long.

Centre Some of the most ornate and neatly
attractive species of brachiopods within the
Paleozoic are represented by the *Spirifer*
group. The specimen shown here is *Spirifer
striatus*, not a particularly common species
from the Carboniferous Limestone of Britain,
but very distinctive and consequently of
immense scientific value. It has acutely
convex valves with a neat triangular fold on
the dorsal surface. The hinge line of the shell
is extensive, and powerful muscles were
required to open and close the two valves.
This specimen is about 15 cm. (6 in.) in
width.

Bottom Brachiopod species range in size from
microscopical to extremely large.
Gigantoproductus giganteus is the largest fossil
brachiopod species, sometimes reaching
38 cm. (15 in.) in width. Such examples are
rare, but smaller specimens are not
uncommonly found in the Carboniferous
deposits of Britain and Ireland. These
brachiopods probably lived on the finer
sediments of the sea-bed which might
explain the extended hinge-line; this would
have increased the surface area of the shell
and would have prevented it from sinking
into the sediments.

Left Lonsdaleia floriformis is a coral which comes from the Carboniferous Limestone of Shropshire, England, but is also found in other parts of the world in rocks of the same age. It is a typical example of the flower-like colonies of these animals called rugose corals. They were important contributors to reef limestones of this period and can sometimes be seen on the polished surface of limestone blocks used to ornament large office buildings. Within the reef, small specialized fish thrived upon the polyps or soft parts of the coral colonies, remaining comparatively free from the predation of sharks and larger fish. Typical fish examples of this sort of fauna can be collected from the Miocene limestones of Monte Bolca in northern Italy (see pages 81 & 84).

Below Many changes were taking place among the aquatic life of the Upper Carboniferous. Most of the swampy forests of the coastal areas had a fauna of freshwater fish and molluscs which had the ability to adapt themselves to the changing environment. With the increased salinity which sometimes occurred in these swampy glades when there was an invasion of the sea, it became necessary for some of the inhabitants to move away. This led certain species of fish to develop the power of breathing the surrounding air as well as absorbing oxygen from the water. Lungfishes, such as *Uronemus*, had become specialized in this way. They also began to adapt themselves to a life on land by using their strengthened fin-bases as rudimentary

limbs. This was a major stage in the development towards the true amphibians. Placoderms and primitive sharks were also present in the brackish or semi-marine waters of the Upper Carboniferous forests. These early genera were highly successful animals and gave rise to many new species, some continuing almost unchanged in form for many hundred thousand years. Their success was also due not only to their obvious carnivorous ferocity but also to their ability to adapt themselves quickly to changing environmental conditions, particularly changes in temperature. Very few fossilized remains of the early placoderms and primitve sharks have been found and we still know little about the evolution of some of the more specialized species.

Right Another type of coral is called tabulate. The specimen here is *Syringopora reticulata* also from the Carboniferous limestone. Unlike *Lonsdaleia* it consists of a reticulate mass of inter-connected tubes with lateral openings where the polyps, or animal parts of the coral, lived. This type of coral was a reef former. Ancestral forms related to this genus are to be found in the tropical coral seas today. Vast colonies eventually build up coral islands which in time become inhabited land areas. Modern examples are provided by the Seychelles and Maldive Islands in the Indian Ocean.

4 The Evolution of Reptiles

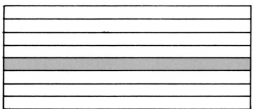

The Permian Period

It is necessary, once again, to look at the progress of geological events. A series of paroxysmal earth movements which began as early as Devonian times continued throughout the Carboniferous and reached a climax during the Permian. These movements resulted in the building of mountain chains which, in time, were to effect the distribution and isolation of animal groups throughout the world.

In Europe a major movement took place and was responsible for building the Hartz mountains in Germany and which had far-reaching effects in other parts of Europe and Britain. Land barriers were formed by former sea-beds which had been thrown up while established land masses just as dramatically disappeared beneath the sea. Hence the necessity for adaptation which, once more, became the only chance for survival for most of the animal populations at the time.

In the southern hemisphere, a vast cooling-down period had resulted in a southern ice-cap which extended almost as far northwards as the present-day position of the tropics. This had quite a marked effect upon the fauna. A temperate zone, with a central sub-tropical and tropical belt, occupied an area much further north than this position, with large areas of warm seas and temperate land masses conducive to the growth and development of reptiles.

Although controversial discussion regarding the origin of the reptiles revolves around one particular line or another, it cannot be denied that by Middle Permian times a vast number of genera and species of reptile had evolved and it looked very much as though many of them were here to stay. The development of *Seymouria* was only the first of a long line of interesting reptiles and reptile-like animals which were later to dominate the whole world of living creatures for quite a long period of time.

During the Upper Carboniferous period, the reptiles established a main stem of evolution and development. This stem, known as the *cotylosaurs*, gave rise to many other types of reptile from Carboniferous to Triassic times. Important groups, such as the chelonians or turtles, probably arose from a special cotylosaur. The Permian genus *Eunotosaurus*, with flattened plate-like ribs, was possibly an early chelonian ancestor which originated in South Africa. Later development of this rather special group occurred during the Triassic period; one branch adapted itself to terrestrial conditions and gave rise to the tortoises.

Other groups of reptiles which developed from the cotylosaurs included various highly specialized species, such as *Pareiosaurus*, a large herbivorous reptile the remains of which have been found in Russia and South Africa. It was a heavy animal with a wide thick-set body and short legs and reached a maximum length of between 8 and 9 feet (2·4–2·7 m.). Its remains have been found in deposits associated with past swamp conditions and it is generally assumed that the wide, spread-out feet with thick toes were developed in order to help the animal stabilize itself under such conditions.

At about the same time that the reptiles were developing in South Africa and parts of Asia, another line was being followed in the New World. Among the deltaic swamps which existed in Texas at this time there was a fierce carnivorous representative of the pelycosaurs, *Dimetrodon*. It was a short, squat form with many of the primitive features seen in earlier species, such as sprawled limbs and slow movement. One character which distinguishes it from

nearly every other reptile was the tall, sail-like dorsal crest. A close relative of *Dimétrodon*, although herbivorous in habit, was *Edaphosaurus*, which lived under similar conditions in the same geographical area. This genus also possessed a tall, dorsal sail-like crest but it was constructed differently, with thicker upright ribs and short side branches. It is not fully understood why these sail-like structures were developed, and suggestions as to their function vary from camouflage to cooling devices.

Apart from its ornate dorsal crest, *Dimetrodon* was an important reptile in the succession of evolutionary forms. Although rather primitive in some of its characters, it possessed tusk-like front teeth and serrated-edged back teeth. These dental features are thought to have been an advance towards the mammal-like characters of some of the later Permian and early Triassic reptiles found in East Africa and parts of Russia and the U.S. The pelycosaurs, therefore, form a link between the ancient primitive reptiles and the more advanced Therapsida with their mammal-like characteristics.

Within this new group were two important subgroups, the Dinocephalia and the Dicynodontia. Of these, the Dinocephalia, which had large heads and were probably about 13 feet (4 m.) in overall length, were represented by two main genera, *Titanosuchus* and *Tapinocephalus*. The first of these was undoubtedly a carnivore, while it seems likely from a study of its teeth that *Tapinocephalus* was herbivorous.

The Dicynodontia, which were closely related to the Dinocephalia, were strictly herbivorous and are represented by a very distinctive genus, *Dicynodon*, which has been found in the Upper Permian rocks of Dvina, Russia, although it also developed later in South Africa and is found in the beds of the Karroo Series of the Triassic period.

Land reptiles were not by any means the only interesting forms of life which were rapidly evolving during the Permian period. In the warm shallow seas which surrounded the Permian continents, lived an abundance of invertebrate life. Corals and reef-forming bryozoa created a protective habitat for the countless species of bivalve molluscs, brachiopods and echinoderms.

Many of the earlier Paleozoic species were extinct by the end of Lower Permian times and were being replaced by genera and species which became highly specialized in both their function and their appearance. Among these was *Prorichthofenia* which closely resembled a cup-coral and lived in a very similar colonial environment. Many fine specimens of this unique genus and numerous other specialized forms have been discovered in the Permian limestones of the Glass Mountains in Texas, U.S. They are often found preserved as silicified replacements of the living shell which can be developed with hydrochloric acid to produce an almost perfect specimen.

Towards the end of the Permian period, a major change occurred. Land levels were once again altering due to many influences, both isostatic and volcanic in origin. This created fresh land barriers and many seas or marine areas became isolated. The climate was, once more, beginning to warm up in the northern hemisphere and vast deserts were formed. These somewhat drastic changes allowed only those species which were more readily adaptable to sudden change in the environment to survive. As the seas increased in salinity, the faunas became dwarfed by the prevailing conditions and many

entire groups of animals died out completely. This led to a rather impoverished though highly specialized animal population on both land and in the sea.

Thus with the end of the Permian we can distinguish the end of the Paleozoic epoch. We are now about to start on a journey through Mesozoic time in which a vast number of very exciting changes occurred. We are about to enter the world of the dinosaur. Before we do this, however, there is still one area of discovery which we have to visit. This is an area which owes its importance to the development of mammal-like features in the reptiles and forges a link between this group and the numerous warm-blooded species which were to develop into the mammals.

Previous page Although some reptiles had shown tremendous advances in their evolutionary style during the Upper Paleozoic Era, others remained as part of the ancient or original stock. The most primitive of these were the cotylosaurs which were represented by a host of diverse species, some large, some small. Among the larger species was *Pareiosaurus* which lived during the late Permian period and reached a length of 3·5 m. (11½ ft). It was a heavy, bulky reptile with thick rough skin and massive, widely spaced limbs on which were five broad toes. It lived mainly in Southern Africa, though had a fairly wide geographical distribution. In spite of its bulk, it was a vegetarian, living on soft plants in swampy areas.

Opposite Acanthodes, seen here as a model, was a member of the oldest known group of true fishes. It represents the spiny, shark-like fishes which ranged from the Devonian to the Permian. This species was found in the Lower Permian of Germany where it is not particularly common, but when found is usually in a good state of preservation. It was not a true shark but had a shark-like dorsal fin on its back. Its long slender body was covered with thick diamond-shaped scales. Most of the specimens found are about 28 cm. (11 in.) in length.

Above right Although brachiopods are a diverse group of marine shellfish still represented in the seas today, the spiny specimens which are illustrated here are examples of an extinct group. They are *Echinauris lateralis* from the Permian Glass Mountains, Texas, U.S. It is probable that the spines were used as stabilizers when the shells were living in the softer sediments on the sea floor. They must have also served to protect the animal from predators. Some of the spines grow to an enormous length but all of those shown here have been broken, probably as a result of the extraction technique when the delicate silicified shells are etched from the calcareous rock in strong acid.

Below right A great many silicified brachiopod specimens representing numerous genera and species have been extracted from calcareous rocks by means of acid techniques. The specimen shown here is of *Prorichtofenia permiana* from the Permian of Hess Canyon, Texas, U.S. As with many brachiopods from this horizon, it has a spiny surface, but this species also develops an elongated conical shell very similar in shape to some of the coral species with which it lived. It also has a fairly complicated sieving device for filtering or sorting its food particles which enter with the water currents through a flap-like valve at the top of the cone. This specimen is 45 mm. (1¾ in.) long.

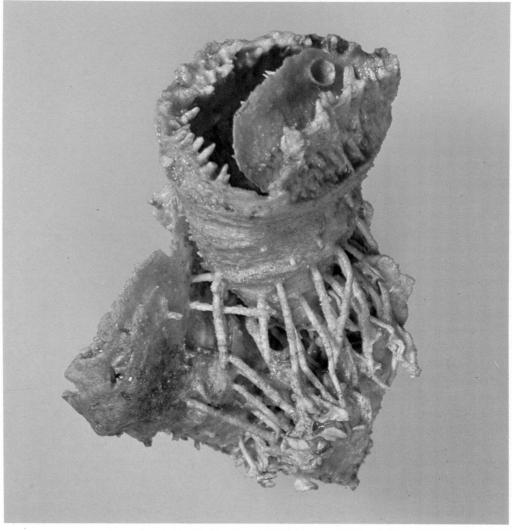

Previous pages The origin of the reptiles has been a subject for discussion and speculation for several years. Some authorities feel that the amphibian *Seymouria baylorensis* provides a link between the reptiles and the amphibians. Reptiles lay their eggs upon dry land and consequently the composition of the shell is different from that of an amphibian egg. Unfortunately, no egg, or part of an egg shell, which can be positively associated with the remains of *Seymouria*, has yet been found and so the argument cannot be concluded. The skeleton shown here is 60 cm. (2 ft) long and comes from the Permian of Texas, U.S.

Right One of the better-known pelycosaurs is the sailback dinosaur *Dimetrodon* which lived during early Permian times in Texas. It is important because it forms a link between the true reptiles of the main stock and the mammal-like reptiles which later developed in South Africa. The outstanding morphological feature which distinguishes this form is the gigantic dorsal fin or sail composed of a framework of elongated spines from the vertebral column, arranged along the middle of the back, which supported a membrane of skin. It is not known for certain for what purpose this fin was provided, but some paleontologists have suggested that it may have served as a cooling device. Its actual size was between 3·5 and 4 m. (11½–13 ft) in length.

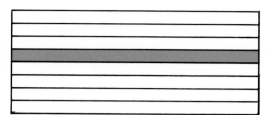

5 The Origin of Mammals

The Triassic Period

The volcanic and other earth-moving activities which had reached a climax in the Permian continued, to a lesser degree, throughout the Triassic period. This was accompanied by more widespread desert conditions and a general raising of climatic temperatures. Already much of the land-locked water in the large lakes of North America and South Africa was drying up and the reptile population was once more obliged to adjust itself to a predominantly terrestrial way of life as a result.

It took several millions of years for this adaptation to take effect, but the genera and species which existed in these continents were well suited to adapting themselves to adverse conditions. It was under such changing circumstances, which prevailed for the better part of the Triassic period, that a major step towards the evolution of the mammals was taken.

In some parts of South Africa today there are deposits of Triassic sandstone in which are found the remains of reptiles with very special features. These features amount to more specialized canine and incisor teeth, for dealing with different types of food, as well as the ordinary reptilian teeth; the limbs supported a lighter skeletal frame, and the skulls were responsible for a smaller proportion of the total body size. A good example of this type of advanced reptile is *Cynognathus* which reached a maximum length of between 7 and 8 feet (2·1−2·4 m.).

The characters of these unusual skeletal remains show such deviation from their original reptilian stock that many arguments and discussions have arisen as to whether they should be classified as ancestral mammals or just advanced mammal-like reptiles. The same arguments could apply to certain living mammals, such as the monotremes *Platypus* and the *Echidna* or Spiny Anteater, which, whilst exhibiting very advanced mammalian features, continue to lay eggs in a reptilian fashion. This is clearly a more primitive method of reproduction. *Cynognathus*, although completely unrelated to the monotremes, might well have been developed along similar lines, exhibiting both mammalian and reptilian characteristics. One thing is clear, however. The Therapsids, of which *Cynognathus* is an excellent example, certainly gave rise to the mammals.

Because the trend towards mammalian development is evident in one group of reptiles, there is no reason to believe that such a tendency occurred in all reptilian groups at this time. In North America, during the late Triassic period, another advance was taking place.

Reptilian evolution had by this time produced more specialized groups of animals, some of which showed a tendency to walk upright on two well developed hind limbs, the feet of which had three large spread-out toes. The forelimbs were smaller and were probably used mainly for balance and, in the case of some carnivorous species, for holding their prey. These rather specialized reptiles can be broadly classified into two major groups, the Saurischia and the Ornithischia, which together go to make up the dinosaurs. They can be distinguished by their pelvic or hip bones; in the case of the Saurischia, these are composed more like those of a lizard, and those of the Ornithischia are built like those of a bird. Further distinguishing characters of these two main groups lie in the type and position of their teeth. The Saurischia nearly always had teeth in the front part of the mouth which were usually conical and sharply pointed, whereas the Ornithischia very often had teeth which were mainly placed along the sides of the jaw and continued to the back of the mouth. The importance of the teeth in the classification of these two main groups will be shown in later chapters.

Most of the reptiles mentioned so far in this chapter have been terrestrial species, but the Triassic seas also had their reptilian faunas. The Plesiosaurs, which were to develop later in the Jurassic period, first made an appearance in the Upper Triassic. It is probable that these long-necked but small headed reptiles were a development from the same main stock as the more spectacular marine genus *Nothosaurus*. However, many paleontologists feel that this genus is more closely related to the Ichthyosaurs, the fish-like reptiles with a short neck and fin-like paddles.

Nothosaurs were fairly widespread in European Triassic seas and remains have been found in north Italy, Switzerland and Germany whilst other representatives have been recorded from Israel, Jordan and as far east as Japan.

In Nevada, U.S., the remains of an early Ichthyosaur representing the Omphalosauridae have been found in the Middle Triassic deposits. This species has a short skull, unlike the more advanced Jurassic species which were to follow, and strong short teeth rather like those of the Placodonts, which were probably used for crushing the shells of molluscs. It is quite likely, therefore, that it was a near-shore dweller which scooped up its food from the offshore mollusc banks.

Throughout the history of evolution there are gaps in our knowledge regarding the exact relationship between one major group of animals and another. Sometimes, even within a major group, it is not at all clear as to just how some animals originated or how a particular feature developed.

The dinosaurs are not without this type of problem and one of the major groups which go to make up this fascinating branch of reptiles, the Sauropoda, has been the subject of some speculation regarding its precise origin. It is generally thought, however, that the Sauropods probably arose during the Triassic period from a group of rather specialized predaceous reptiles known as the Plateosaurs. The remains of large numbers of this group have been found in southern Germany and their skeletons have been carefully studied.

As a result the skeletons have been reconstructed and these reveal large, somewhat awkwardly built reptiles with large limbs and small heads. Although they were probably bipedal in locomotion, they clearly suggest a link with the bigger quadrupedal Sauropods which were later to dominate the world of the dinosaur.

During the time that the more spectacular animals were developing on land and in the sea, there was also a continuation in the development and evolution of the major marine invertebrates, and also molluscs, brachiopods, corals, and many other primitive forms of life which were most important in the earlier Paleozoic times.

Some flourished whilst other species became extinct, for the Triassic was a time of many major climatic changes. The extra salinity of the seas produced its own problems and many species could not adapt to such conditions. The survivors, however, went on into the Jurassic when they developed from strength to strength.

Previous page A very important group of mammal-like reptiles received considerable publicity in 1969 when a specimen of the genus *Lystrosaurus* was discovered in the Antarctic regions by a group of American paleontologists. The discovery gave the scientists new data on ancient climates and more reasons for speculation on possible positions of continental masses which may have drifted apart after Triassic times. The genus *Lystrosaurus* is rare, but a member of the same faunal group, *Thrinaxodon*, is less rare and is known to occur with *Lystrosaurus* in South Africa and parts of India. The model shown here is of a *Thrinaxodon*

specimen from South Africa and is about 0·5 m. (19⅝ in.) in length.

Below Megazostrodon was an important shrew-like animal no bigger than the domestic mouse. The genus occurred in late Triassic and early Jurassic times in Lesotho, southern Africa, and was one of the earliest ancestors of the mammals. Its long, thin body, weasel-like head and long, tapering tail were covered in coarse hair. The relative length of its hind legs and the slim feet and slender toes indicate that it was a fast-moving animal which probably spent its time in the thick bushy undergrowth.

Bottom The early ancestors of the dinosaurs were comparatively small, lightly built reptiles which relied very largely upon their well-developed hind limbs for survival. It was due to this feature that they were able to move swiftly away from predators. *Ornithosuchus*, represented here by a reconstructed model, was typical of this type of reptile which lived during the Triassic period. It is seen in an almost upright position with its short, almost useless, forelimbs and the long tapering tail which it used in order to balance itself. It reached a height of approximately 1 m. (3¼ ft).

Left Many species of amphibia developed during the Triassic period and were distributed all over the world. The model pictured here is a reconstruction of *Paracyclotosaurus* from the Upper Triassic of New South Wales, Australia. It was a large species, reaching a length of approximately 3 m. (10 ft). A particular feature of this amphibian was its enormous, wide head. From the bones of its skull and other parts of its skeleton, paleontologists are able to deduce that this type of amphibian had previously established itself on land and had now returned to life in the freshwater lakes and pools.

Below The aquatic reptile *Nothosaur* had many points in common with the plesiosaurs and may well have represented an earlier group which gave rise to them. It had the characteristic long neck and broad flattened skull, although the head was probably more elongate than most species of plesiosaur. The legs, however, were adapted for walking short distances, unlike the paddles of the plesiosaurs, but the feet may well have had webbed toes to assist in locomotion through the water in which it spent a great deal of its time. This skeleton is from the Triassic rocks of Austria. It is approximately 23 cm. (9 in.) long.

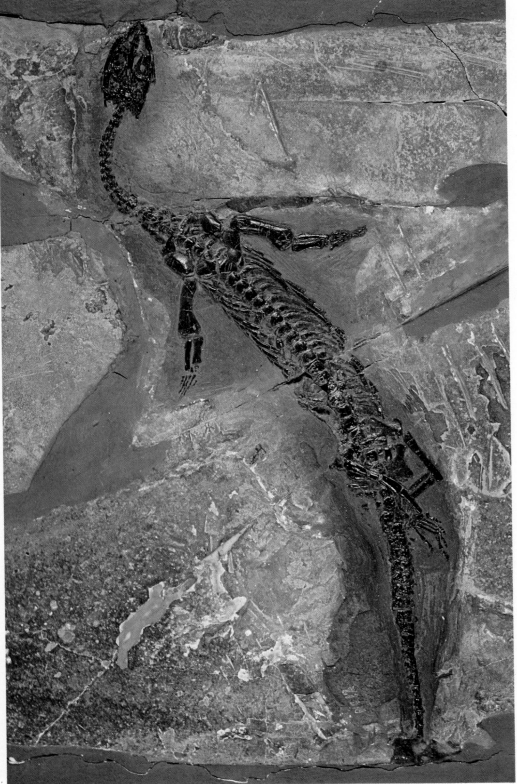

Opposite In South Africa a number of skulls of *Cynognathus* have been found. These are important finds for they allow closer study and comparison of this strange reptile with the skulls of more conventional forms. From these studies we have learned that *Cynognathus* and similar genera have skull bones, teeth and jaw assemblages very much like those of some primitive mammals. It is quite likely that this genus was only a stepping-stone towards mammalian development.

Opposite below It is not always possible to show distinct dioramas for both freshwater and marine life. This Triassic scene is of a composite fresh and saltwater fauna. The placoderms which originated in the early Paleozoic Era were now extinct and were replaced by other genera, such as the freshwater shark *Lissodus*. Meanwhile, in the Triassic oceans, flying-fish represented by *Thoracopterus* had developed.

Below During the Triassic times changes occurred which were due either to sudden earth movements or to the drying up of lakes by drought conditions. The freshwater fish shown here are from the Upper Triassic of South Africa and belong to the species *Seminotus capensis*. It seems that they may have been part of a large shoal suddenly overcome by a catastrophe of this description. Similar species of *Seminotus* also occur in Triassic rocks of Arizona and reach approximately 35 cm. (13¾ in.) in length. They represent the group called Holosteans and are characterized by their deep bodies and strong, peg-like teeth which they used for crushing bivalve molluscs and other invertebrates.

Following pages The ruling reptiles or archosaurs were the main reptilian stream from which various specialized branches have sprung during the course of geological time. *Euparkeria*, shown in the model, was a carnivorous lizard-like reptile from the Triassic period which, although very closely bound to the main reptilian stream, shows some specialization in having elongated back legs. It is likely that the reptile moved around swiftly on just the two back legs when in pursuit of its prey. It was a fairly small animal, standing no more than 1.5 m. (5 ft) high.

6 The Evolution of the Dinosaurs

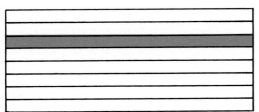

The Jurassic Period

After the turbulent earth movements of the Permian and Triassic periods, there followed a fairly long period of quieter geological times over the entire world. The steady deposition of sedimentary rocks continued for many more millions of years without any major catastrophic breaks. Thus we note a burst of life on land and in the seas during Jurassic times which lasted for approximately forty-six million years. It was during this time that perhaps some of the most exciting and dramatic changes occurred within the animal populations, especially those of the reptiles. In order that we should appreciate the extent of these vast changes, we must be prepared to transport ourselves metaphorically from one continent to another and from one environment to an altogether different one.

The trend to a more aquatic existence among the reptiles which began in the Triassic period continued unabated in Europe into the early and middle Jurassic, finally reaching a climax in late Jurassic and early Cretaceous seas. By this time many of the species originating from Europe had spread to the North American, Asian and Australian continents.

Many fine examples of enormous plesiosaurs, with barrel-shaped bodies and elongated necks, were found in Liassic deposits of the Lower Jurassic strata in England during the latter part of the last century and the beginning of this. They averaged from 12 to 20 feet (3·7–6·1 m.) in length, although some species reached the immense size of 40 feet (12·2 m.) from head to tail. In spite of their enormous length and comparatively bulky bodies, they had small flattened heads equipped with rows of sharp conical teeth which were used to grasp the cuttlefish, ammonites, and numerous fish upon which they fed. They propelled themselves across the surface of the water like a partly submerged submarine, using the vast paddle-like limbs as though they were oars.

Living alongside the plesiosaurs, sharing their food and possibly attacking them from time to time, were the ichthyosaurs, large fish-like reptiles with erect dorsal fins, shark-like tails, and long snouts with rows of very sharp teeth. Although not as large as plesiosaurs, they could probably move through the water at a very much faster rate, snatching their food of fish and squid as they went.

At about the same time as this re-adaptation to the aquatic way of life, the larger land reptiles evolved to reach a most dramatic stage. The sauropods, represented in England and the European continent by Cetiosaurus, were also to be found in North America where they reached enormous proportions. The best-known examples of this group are Diplodocus and Apatosaurus (Brontosaurus) which developed in the marsh plains of the western U.S. In spite of their immense size, these reptiles were strictly vegetarian and probably lived on the numerous cycads and tree-ferns which abounded at that time. They are often illustrated as half immersed in water of swampy land areas, but some authorities believe that they inhabited the firmer land. One point in favour of their having dwelt in a watery environment would be that some species had a nostril or nose-channel at the top of their skulls, sometimes between the eyes, so that they could become almost entirely submerged in water but still be able to see and breathe. The largest of the sauropods was Diplodocus, which weighed an estimated 30 tons and achieved a length of 87 feet (26·5 m.). It had a comparatively tiny skull at the end of an elongated neck, and a huge barrel-shaped body, vast elephantine legs, and a long thin tapering tail.

Not far from the habitat of these giants, a more slender but heavily armoured reptile of the Ornithischia group also browsed on the ferny plains. This comparatively docile creature was, nevertheless, quite a match for its carnivorous predators since it had a double row of spear-like plates upon its back, and a long spiky tail. At the base of its vertebral column, it had a secondary nerve centre for controlling its movements, and a tiny brain for thought processes in its small skull.

Having successfully established themselves on land and in the sea, it seemed only a matter of time before the reptiles overcame the forces of gravity and conquered the air. This was achieved during the late Jurassic times in two successful but completely unrelated ways.

First, as a result of adaptation to a changing environment, a group of small land-based reptiles had advanced to the point of flight by modifying their bodies so that they became light framed. Then, by the exaggerated extension of one finger joint of the forelimbs and the stretching of membranes to a point of attachment at the hind-limbs, the reptiles became able to glide for short distances, probably scooping up food from the sea before alighting upon the cliffs where they dwelt. This sort of primitive flight remained a highly successful means of locomotion for many millions of years. Several species were developed: Rhamphorhynchus, with a long tail and a thin, pointed head, and Pterodactylus, with a short tail and elongated beak-like snout were among the better known of the Jurassic forms, but Pteranodon was a Cretaceous species which was probably more successful and was a lot fiercer.

The second way in which the reptiles conquered the air was much more dramatic and was to last for many millions of years. It is, in fact, still going on today for the birds are a living witness to the adaptation to a new environment undertaken during the Upper Jurassic period by specially equipped reptiles. Although living at the same time as the pterosaurs, they were quite unrelated but had, in the same way, responded to the stimulus of the air, presumably in the search of food. We can only imagine that this highly adaptable group produced a rather special insulated scale which we now call a feather. This, together with a light hollow-boned skeleton, and powerful hind-limbs, gave it all the necessary motive power for a take off.

It was, therefore, only a question of time before the short bursts of forward motion gave way to a soaring glide and thence to movement of the wings to gain height. It all sounds so very simple but it happened over a period of some millions of years and must have taken quite a long time for each stage to progress. The number of skeletal remains of this first bird, the Archaeopteryx, amount to only six or seven. They all come from the Solenhofen Stone of the Bavarian Upper Jurassic in Germany. They had a typical reptilian head with sharp teeth, and a spread of feathery wings and a long plume of a tail for balance. At the end of each wing was a three-toed claw.

While all this activity proceeded inside the reptilian world, the life beneath the seas continued. The rise in temperature towards the end of Jurassic times had encouraged the growth of the cephalopods, the ammonites and squid, which are so characteristic of the limestones of the Lias, Oolites and Upper Jurassic zones. Many of these species reached enormous proportions, especially towards the end of the

Jurassic, and Portlandian ammonites from the Portland Stone of Dorset, England, sometimes reach 4 feet (1·2 m.) in diameter. Ammonites are particularly useful to geologists because they had a very fast rate of evolution and were very sensitive to a change of conditions. This enables the geologist to mark out horizons within the strata, using the different species of ammonite as zonal indicators.

This was also a good time for the other molluscs, the bivalves and gastropods. Many species of molluscs were produced during the Jurassic and these are also used, in a less specialized way, to mark the horizons and zones within the stratigraphical layers.

Previous page Archaeopteryx, the first bird, was developed from a very specialized reptile. From the very few specimens which have been found in the Upper Jurassic limestone from Germany, it is possible to identify many purely reptilian features. This unique species had a slim body with a long tapering tail, two lightly built wings which ended in three reptile-like claws. The rounded skull had a set of forty-eight small sharp teeth in a slender pointed beak, and large rounded eye sockets. Two muscular legs with efficient perching feet supported the lightly built frame. The body was covered in reptilian scales but the wings and tail were provided with long feathers. Impressions of the feathers have been found with the skeletons. It is probable that *Archaeopteryx* was capable only of launching itself from trees and gliding for short distances, rather than of sustained flight. The best preserved specimens are about 46 cm. (18 in.) long.

Opposite Unlike the contemporary *Ichthyosaur* (see page 57), the *Plesiosaur* was long-necked and rather barrel-shaped, but equally adapted to its marine environment. Most species of *Plesiosaur* had oval, streamlined bodies supported by strong flattened ribs. They had long tapering tails free of any fins. The model, shown here catching a small fish, represents a species called *Macroplata*, which lived in Jurassic times and attained a length of approximately 4 m. (13 ft). Some species, however, were extremely large, reaching around 16 m. (52½ ft) in total length with a

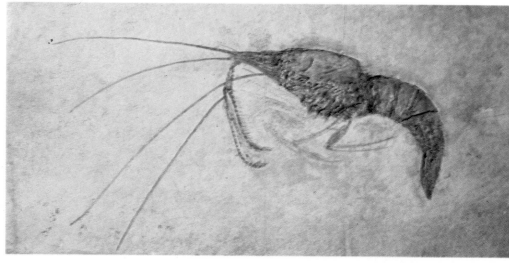

head of anything up to 3 m. (10 ft) long. Strong, comparatively long, paddles allowed the reptile to row itself through the water at speed, although it could never have been so manoeuverable as the fish-like *Ichthyosaur*.

Below The *Pliosaur* was a curious marine reptile, almost midway between a *Plesiosaur* and an *Ichthyosaur* in morphological features. It had a well developed elongate-oval body with large paddles like those of a *Plesiosaur*, but the head was narrow and had an elongated snout, rather like that of the *Ichthyosaur*, with rows of sharp conical teeth. It was not a particularly successful reptile and lived for a comparatively short span in the Upper Jurassic seas. The skull shown here is approximately 1 m. (3¼ ft) in length.

Above The decapods are the most diverse of all crustacea and include crabs, lobsters, prawns and shrimps. *Aeger insignis*, a shrimp, is shown here in a particularly good state of preservation. It comes from the Upper Jurassic limestone of Solenhofen, Germany.

Bottom The excitement aroused by the appearance of a living coelacanth fish was understandable, but such examples of living fossils are not as rare as is generally thought. Although the actual species is extinct, *Chondrosteus* which lived in early Jurassic times in Europe, is represented in modern seas by the comparatively modern American sturgeon, not all that different from the fossil. The model shown here is about 40 cm. (15¾ in.) long.

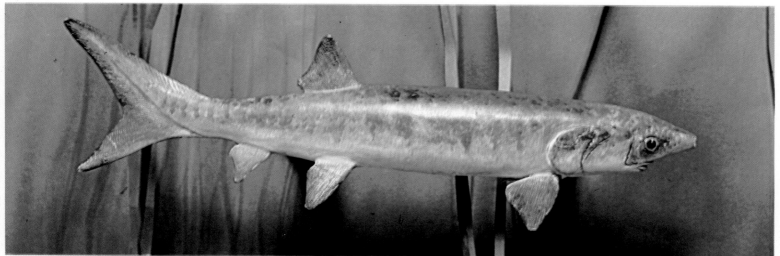

Previous pages Throughout the history of evolution there are times when, for some seemingly unexplainable reason, the balance of species within a given environment appears to change suddenly. In fact, we know that these changes occurred over many thousands, perhaps millions of years. However, the fauna often shows a marked change which can sometimes serve as a very convenient position to divide and record the stratigraphical sequences. The difference between the fish faunas of the Lower Jurassic and the Upper Jurassic is quite marked. In strata which have been deposited under marine conditions, we find an increase in the primitive bony fish at the beginning of the Upper Jurassic, which is a departure from the older forms which had their origins in the very early Paleozoic and were still quite recognizable into the Triassic and, in some cases, the Lower Jurassic. In Liassic Seas, represented in the diorama, *Acidorhynchus*, *Hybodus*, *Dapedius* and *Chondrosteus* all show some of the more primitive features of their early ancestry.

Below left It has already been said that the inside of an ammonite shell consists of a series of chambers separated by walls or septa which mark the limits of each chamber. The line of junction between each septum and the inner surface of the shell wall is called a suture-line. These lines have a distinct pattern which can be interpreted and classified according to the genus of the ammonite. The suture-lines themselves are placed into certain categories. The one shown on the Jurassic ammonite here is a *ceratitic* suture-line. The specimen is approximately 9 cm. (3½ in.) in diameter.

Below centre The *Nautilus* which lives in some modern seas (see page 8) has changed very little throughout its evolutionary history. The specimen shown here is of a sectioned fossil *Cenoceras* species from the Liassic limestone of the Dorset coast. It again illustrates the separate living compartments or chambers and the siphuncle passing through the centre of each septum which is used to control the buoyancy of the animal. In the ammonoids, the suture line is arranged in an elaborate pattern, but in *Nautilus* and other nautiloids, the suture line is plainer.

Opposite right Although this sea-urchin specimen appears to be of regular shape, it is, nevertheless, a member of the subclass Irregularia and is a very important horizon marker within the Middle Jurassic rocks of Europe. This specimen is *Clypeus ploti* from the Inferior Oolite of Birdlip Hill, Gloucestershire, England, and is about 7 cm. (2¾ in.) in diameter.

Opposite bottom By the end of the Jurassic period, the climate had changed and the rate of deposition within the seas had also altered. Not surprisingly, such changes brought about certain faunal differences. *Heterodontus* and *Pholidophorus* and many sharks, skates and rays were replacing the faunas of the Liassic and Triassic seas. Thus it

is that we are able to draw a fairly definite distinction between the faunal groups of the Liassic and Upper Jurassic periods.

Below and bottom Ichthyosaur means fish-reptile and it is easy to understand why this aquatic animal received such a name. It serves as a perfect example of the way in which animals adapt themselves to their environment. Superficially, this genus is very fish-like, with its short neck and elongated snout, large round mobile eyes, dorsal fin and fin-like paddles. It also had a fish-like caudal or tail fin which resembled that of a modern shark. Everything about this reptile is designed to give it an easy passage through the water in which it lived. The model shown here is of an *Ophthalmosaurus* which averaged 2–3 m.

(6½–10 ft) in length. The skeleton, which is of another species, shows the outline of the actual skin of the reptile still perfectly preserved and allows us to trace the contours of the body and apply them directly to the bony framework within. Both forms lived during the Jurassic period and were fairly common predators around the shallow coastal areas of Europe. Examples of *Ichthyosaur* skeletons have been found showing what appears to be the skeletal remains of unborn young within the abdominal cavity. It is assumed, therefore, that this particular reptile retained its eggs within the body until they were mature embryos, and, when they were finally discharged from the body of the mother, they would almost immediately hatch.

Left The squid belongs to a group of cephalopod molluscs known as dibranchiates which are close relatives of the octopus. Among the numerous species in this group is the common cuttlefish *Sepia officinalis*, a model of which is shown here to illustrate what the bullet-shaped belemnite, seen with it, was like when alive. The fossil equivalent of the squid is usually found only in this form – the main support for the soft parts of the animal which have long since decomposed. Belemnites are comparatively common fossils in clays and some limestones in the Mesozoic Era, particularly in places like Lyme Regis, England, where they occur in the Liassic deposits. They are particularly useful horizon markers because, like the ammonites, they evolved very quickly.

Centre left The famous Upper Jurassic limestone from Solenhofen in Germany has provided many fine specimens of fossil animals. Under any circumstances, insects rarely become fossilized as they most probably disintegrate quickly after death. Occasionally a fine example comes to light in the Solenhofen limestone. This dragonfly, *Cymatophlebia longialata*, illustrates just how perfectly such insects have been preserved.

Below left Apart from the winged insects which arose during the late Paleozoic Era, the conquest of the air did not occur until the Upper Jurassic period. Two completely different lines of reptiles mastered the art of flight in completely different ways. One of these was developed by the frail, lightly built *Rhamphorhynchus*, shown here, with a long tail, toothed beak and extended finger bone from which was stretched a thin membrane like the wing of a bat. *Rhamphorhynchus* was about 61 cm. (2 ft) long and was capable of short bursts of flight. Most of the skeletal remains have been found near shore lines and lakesides and it is assumed that it fed on fish which it scooped up with its long beak. The short, backward bent claws on the hind limbs probably meant that the reptile hung upside down from cliffs or trees when resting.

Opposite above By mammalian standards none of the larger dinosaurs could be described as attractive animals, but one of the oddest looking members of this group must surely have been the *Stegosaurus* with its double row of pointed protective armour plates surmounting its back. This comparatively docile herbivore must have presented a difficult problem for its predators.

Opposite below In some of the rocks near Canyon City in Colorado, U.S. a number of specimens of the fierce predaceous carnivore *Ceratosaurus* have been found. At one time, the sort of scene depicted here must have been commonplace among the reptile communities of the northwestern states. Most skeletons that have been found are about 6·5–7 m. (18½–22¾ ft) in length. The development of the hind limb bones suggests that the reptile spent a lot of its time in an upright position, similar to the megalosaurs, and was probably capable of moving fairly rapidly. It had an extensive hinge joint to its jaw, far back in the skull, which made it possible to open its mouth very widely. It was thus able to consume vast chunks of flesh from its victims in a very short time and consequently became less vulnerable to attack from other predators. They lived during the middle part of the Jurassic.

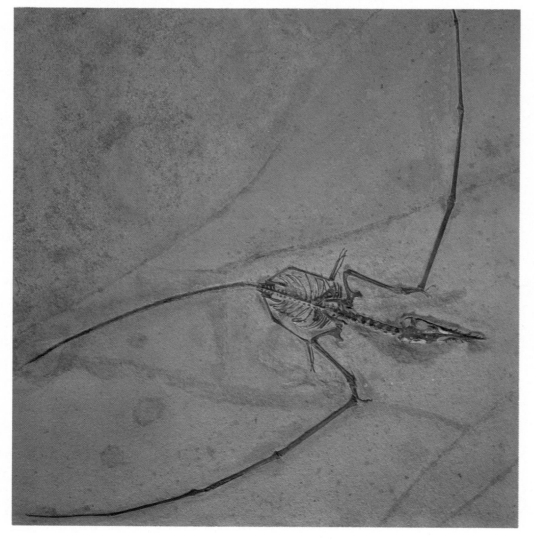

Previous pages *Apatosaurus*, more commonly
known as *Brontosaurus*, was an immense
herbivorous reptile which lived during the
late Jurassic period in the western U.S. The
model is depicted here on hard ground, but
is more likely to have spent a great
proportion of its time in swamp land partly
immersed in water. Such an enormous
animal would have required a vast amount of
vegetable matter to maintain it, and this is
yet another reason for believing that it had
a watery habitat where large quantities of
water weed would have been readily
available. The most commonly quoted
member of this group is perhaps *Diplodocus*
which was 30 m. (98½ ft) long and must
have weighed around 30 tons.

Right Scenes of carnage, such as the one
shown in this diorama, must have occurred
quite frequently in Upper Jurassic times.
Allosaurs are shown here, tearing the flesh of
a freshly killed herbivorous giant, probably
a *Brontosaurus* or related form. The presence
of fierce carnivorous species must have
contributed very considerably to the
extinction of the larger herbivores.

7 The Decline of the Dinosaurs

The Cretaceous Period

The story of the dinosaurs, which began in the Triassic period and developed in the Jurassic, was now reaching a climax and then a decline in the Cretaceous period. By comparison with the immense burst of activity which was to follow, the Jurassic reptilian scene was no more than a curtain raiser in the dramatization of this fascinating saga.

In Europe, however, a vast transgression of the oceans, which started in the late Jurassic period, continued throughout the Cretaceous. This oceanic transgression not only produced new sea areas from older land masses, but also brought the eastern faunas established in Jurassic seas into contact with the western faunal provinces. The same was true of some southern marine faunal groups which were established along the southern reaches of the Tethyan Ocean, an ancient sea which spread outwards from an area roughly occupied by the Mediterranean today.

At the same time, the resultant change in land levels produced barriers where none had existed before. One of these barriers extended across the northern part of the European continent and resulted in the appearance of a vast system of lakes and rivers spreading from southern England, across to north Germany, and down in an area south of the Paris basin in France. Within this vast deltaic system, the lakes were surrounded by lush vegetation of trees and ferns making an ideal setting for the development of reptilian species.

By the beginning of what geologists refer to as Wealden times, the primitive Ornithischian or 'bird-hipped' dinosaurs had developed in southern England and northern Europe. An English species, *Hypsilophodon*, was a comparatively small reptile with a long tapering tail, long muscular hind limbs and shorter, less developed forelimbs. It spent a lot of its time among the branches of trees and larger shrubs, and possibly grew to a length of more than 5 or 6 feet (1·5–2·1 m.) but was only 2 to 3 feet (0·6–0·9 m.) high.

Closely related, but considerably larger, was another herbivore, *Iguanodon*, which has the distinction of being one of the earliest known dinosaurs. Many fine examples of this interesting species have been found in southern England, but a most extraordinary collection of over thirty skeletons was made from deposits of Wealden age at Bernissart in Belgium in 1878. Such a collection may have been the result of some catastrophe, or the deposition of the remains of dead animals by the action of a river. The Bernissart species is probably the largest of the iguanodons and stood 16 feet (4·9 m.) in height and was over 31 feet (9·5 m.) in length.

Iguanodon is characterized by its elongate horse-like head with large oval nostrils, lateral rows of large, flattened grinding teeth and bony spur-like thumbs on its five-fingered forelimbs. Although more commonly found in Europe, remains of this reptile have been discovered in north Africa and a closely related species was found in the Cretaceous deposits of New Jersey, U.S., in 1856. This is *Anatosaurus*, one of the 'duck-reptiles' so called because of its shovel-shaped mouth which in some ways resembles the beak of a duck.

Similar conditions prevailed about this time on the American continent. The regression of the seas left large land areas exposed. The appearance of *Anatosaurus* was only the beginning of a changing scene of both herbivorous and carnivorous dinosaurs which wandered across the western plains of this vast land area. There were fine examples of the herbivorous armoured dinosaurs also living on the plains. Some of these were covered with hard bony plates of specialized skin or had rows of spikes or horns on their heads.

Perhaps the best-known group of the armoured dinosaurs is the Ceratopsia which had their origin in the east, where skeletal remains of *Protoceratops*, complete with clutches of eggs, have been discovered in the Gobi Desert of Mongolia. Their armour consisted of a bony extension of the beaky skull, covering most of the more vulnerable parts of the upper neck. It was arranged like a collar or frill, almost detached from the rest of the body.

This group is represented in North America by *Triceratops* which, in addition to the bony frill of neck armour, also had a short nasal horn, and two more highly developed and longer horns situated above the orbits or eye sockets in the skull. It reached a maximum length of 30 feet (9·2 m.), of which 7 feet (2·1 m.) comprised the head, and it must have somewhat resembled the present-day rhinoceros, although they are in no way related.

There were other armoured dinosaurs at this time, some of which were a further development of some earlier forms from the Jurassic period. One of the most spectacular armour-plated dinosaurs was *Ankylosaurus* which had a crouching appearance and was low on the ground. It had rows of hard scutes, like those of a crocodile, upon a broad back, and a heavily plated head-shield. Sharp, pointed and specially adapted plates protruded laterally from head to tail. On the end of the tail was a cluster of bones or bony plates arranged in a club-like mass. It has been suggested that this structure was used as a weapon of defence, although it is difficult to imagine what possible use it could have been against the prevalent predators of the time.

Towards the end of the Cretaceous period, the reptilian scene in North America and parts of north Asia and Russia was dominated by the appearance of fierce bipedal gigantic saurischian dinosaurs; the best known of these occurred in the western states of the U.S. and was called *Tyrannosaurus rex*. This enormous reptile had a massive head complete with large sharp conical teeth. It stood 16 feet (4·9 m.) off the ground and was about 50 feet (15·3 m.) in total length, the head accounting for about 4 to 5 feet (1·2–1·5 m.) of the total. Two ridiculously small, two-toed forelimbs hung from its massive shoulders and it remained erect on two powerfully mascular hind limbs. Although terrifyingly fierce and horribly grotesque, it probably moved comparatively slowly, for this massive beast must have weighed anything from 7 to 10 tons. Unlike the *Iguanodon* and the armoured dinosaurs mentioned previously, *Tyrannosaurus* was exclusively a carnivore or flesh-eater and preyed upon the less fierce grazing reptiles which lived alongside it.

Many paleontologists have put forward theories regarding the decline and eventual extinction of the dinosaurs, and certainly there seems room among the numerous hypotheses for the suggestion that fierce predators, such as the tyrannosaurs, may well have accounted for their demise. However, the dinosaurs had already greatly declined before the predominance of such powerful carnivores. It may have been that a changing climate over many thousands of years had altered the balance of vegetation to such an extent that many varieties of plant had ceased to flourish or had died out altogether. Denied their usual grazing, the herbivorous dinosaurs would have had difficulty in adjusting to a new environment; consequently they would have starved, or declined to an extent that they would have become easy targets for the predaceous tyrants which succeeded them.

Whichever theory holds good, it is a fact that the dinosaurs became extinct at the end of the Cretaceous period. By this time, however, they had been partially replaced by a fast developing group of animals which had made their debut in the Triassic times. These were the earliest mammals and we shall see in the next chapter how they were to develop and dominate, and then to become the most important group of animals of all time.

Meanwhile, the vast transgressive seas, which affected Europe and parts of the North American continent and brought about the isolation of terrestrial animal groups, were also fostering the invertebrates in warm shallow waters around the continental shores.

Numerous species of ammonites, belemnites, bivalve molluscs, corals and brachiopods flourished until well into the late Cretaceous. At the beginning of the deposition of the Chalk, however, a marked change occurred. Whether or not the temperature began to rise steeply at about this time is not at all certain. Desert conditions prevailed throughout most of the Upper Cretaceous times and any sudden increase in temperature could well have accounted for the comparatively noticeable changes. Many of the species which had developed and flourished in the earlier part of the period were becoming extinct, until only a very few remained to give us some idea of the contemporary conditions on the sea floor.

Previous page Tyrannosaurus rex was the largest carnivorous dinosaur, reaching a length of over 15·5 m. (51 ft) and standing over 6 m. (19¾ ft) high. It was a fierce reptile which lived on the plains of north America, parts of Asia and the U.S.S.R. during the late Cretaceous period. Although by no means common, remains of this giant show the extraordinarily large head in proportion to the body, the sharp dagger-like teeth, approximately 13 cm. (5 in.) long, and the small, two-toed, forelimbs which could only have been of use in holding its prey. With related forms, such as *Gorgosaurus*, the *Tyrannosaurus* was a further development of *Allosaurus* which was a more lightly built dinosaur from the Upper Jurassic period. The model shown here in an ideal sub-tropical setting gives a rough idea of the extraordinary spectacle that this enormous reptile must have presented.

Opposite The ceratopsian reptiles are those curious looking giants with horns upon their skulls and extensive bony protective frills around their necks. We can see *Triceratops* which had three horns on its head on the following pages, but this model shows an even more ornate array of spikes in addition to a central nasal horn on its beak-like snout. The spikes are in fact an adaptation of the bony neck frill and must have served as a means of defence in skirmishes with carnivorous predators. This species is *Styracosaurus* from the Upper Cretaceous of North America.

Below and bottom As in the case of *Inoceramus*, the bivalve illustrated here, *Spondylus spinosus*, can be distinguished by its external shell ornament. In this case the ribs or costae are longitudinal, radiating from the umbo (apex) near the hinge. In addition, there are numerous long spines attached to the surface of the shell which probably helped to stabilize the animal on the sea-bed and would also serve as camouflage. The spines themselves probably discouraged likely predators. This genus is still represented in

the seas today, and the specimen shown here agape is the thorny rock oyster, *Spondylus petroselineum*.

Below During the early part of the Cretaceous period, a great area of river deltas and lakes covered southeast England, parts of north Germany and northern France, stretching from Dorset in the west to the eastern part of the Paris basin. The whole of the area was covered in small shrubs, ferns and trees which provided excellent conditions for the development of reptilian life. One of the largest reptiles to inhabit this area was *Iguanodon*, an enormous herbivore or plant-eater which spent much of its time around the sides of the lakes. It walked upright on two highly developed hind limbs. On its front legs it had a sort of horny spike at the end of the fifth digit or thumb, seen in the model, which is characteristic of the species.

Above It is not often that the claws of crustaceans are found in the chalk but, when they are, they are very often well preserved. The specimen shown here is the claw of a lobster *Palaeastacus sussexensis* from the Lower Chalk of Burham, Kent, England. The walking legs and the smaller appendages, which the lobster uses for swimming, are very seldom preserved in the fossils.

Previous pages Triceratops was a large, three-horned herbivorous dinosaur which also lived in the Upper Cretaceous period. Its skeletal remains have been found only from the north American continent, but it was developed from animals such as *Protoceratops* which originated in late Jurassic times in Asia. The largest member of this genus reached a length of approximately 8 m. (26 ft) and probably weighed about 5 tons. The characteristic neck frill was a protective bony collar produced by an extension of the skull bones. Together with the sharp horns on its head, this frill helped to ward off the attacks from its more ferocious predators.

Opposite top left The Mesozoic Era provided just the right conditions for the growth and survival of numerous invertebrates. Among the most spectacular of these were the ammonites, many of which reached their evolutionary climax in Jurassic seas. Some forms survived into the Cretaceous where they lived on, producing another short burst of development and numerous species followed. These species evolved quickly and have proved of tremendous use as fossil horizon markers and zonal indicators. The specimen figured here is of *Hoplites dentatus* from the Middle Albian clays of Folkestone, Kent, England and gives its name to the *dentatus* Zone. Specimens vary in size but a good average would be about 5 or 6 cm. (2–2$\frac{3}{8}$ in.) in diameter.

Opposite top centre Some of the longest ranging and better organized members of the articulate brachiopods are the rhynchonelloidea. The group pictured here belong to *Cyclothyris difformis* which were fairly commonly associated with shallow water faunas of the sea-bed communities during the Cretaceous period. This particular species occurred in the Cenomanian seas off the coasts of France, north Germany and England. These specimens are approximately 25 mm. (1 in.) long and 32 mm. (1$\frac{1}{4}$ in.) wide.

Opposite top right Within the sandy beds of the Lower Cretaceous rocks are many fossil bivalve molluscs. Some of these are like large flat oysters while others are smaller cockle shells, similar to those found on our beaches today. An interesting species, looking like a very elongate mussel shell, is *Gervillella sublanceolata* seen here in a cluster. It is not common, but is found from time to time in the Lower Greensand beds of the Isle of Wight, England, and other places in the Weald area of southeast England. The individual specimens are about 23 cm. (9 in.) in length.

Below At the end of the Jurassic period, world temperatures began to rise again and by the middle and end of the Cretaceous there was a return to the very hot desert conditions which had existed during the Triassic period. The result of this rise in land temperature meant that there was a corresponding increase in sea temperature. The whole of the Chalk seas were in fact warm and shallow, probably not exceeding 183 m. (100 fathoms) in depth. By this time

the fringe-finned fish and coelacanths had dwindled to near extinction. Their ecological niche was now being occupied by more and more bony fish similar to herrings and eels, which became numerous and varied.

Opposite below right Although most ammonite species appear as flat coiled shells, there are some species which changed this regular pattern of growth and began to uncoil. Some became almost straight, while others uncoiled themselves sideways and looked rather like a large gastropod or sea-snail. The specimen here is one of these uncoiled forms called *Bostrychoceras polyplocum* from the Upper Cretaceous of Haldem in Westphalia, Germany. It is approximately 27 cm. (10$\frac{5}{8}$ in.) long. It is

quite probable that the rather bizarre shapes which were assumed by these ammonites contributed very largely to the extinction of some of the species.

Left The group of conical objects seen here might easily be mistaken for corals but they are, in fact, bivalve molluscs which have become specialized. They belong to the group known as rudists, represented here by the genus *Hippurites* from the Upper Cretaceous of France. They consist of a deep cone-shaped right valve, and a thinner lid-like left valve, which are hinged together with peg-like teeth and deep sockets. The group tends to form small reefs in fairly shallow water, attaching themselves to rocks on the sea-bed. The specimens often attain quite massive proportions, although those figured here are only 16 cm. (6¼ in.) in height.

Centre Among the numerous representatives

of the Echinodermata are the sea-urchins or Echinoids which are delicate animals enclosed within a spiny calcareous test (case) made up of a series of interlocking plates. These animals conveniently fall into two distinct subclasses. The first of these is the Regularia which are often perfectly circular in general outline and range from almost spheroidal to broadly barrel-shaped. They have club-like spines which are attached to the plates of the test by means of strong muscles and ball-and-socket joints. The second subclass, the Irregularia, range in outline from sub-circular or oval to heart-shaped and are covered in fine, almost needle sharp, spines. The specimen figured here is the regular echinoid *Stereocidaris sceptrifera* from the Upper Chalk of Kent, England, but similar species occur in the Mesozoic Era from practically every continent. The diameter of the specimen shown here is about 5 cm. (2 in.).

Below left We have seen that the pterosaurs began their strange airborne existence during the Jurassic period with such forms as the long-tailed *Rhamphorhynchus*. By the end of the Cretaceous, the group was represented by the enormous *Pteranodon* which had a wingspan of up to 8 m. (26¼ ft). It was short tailed and had a long, slender beak-like mouth which was toothless, unlike the Jurassic species which had rows of small sharp teeth along the jaws. Upon its head was a tall bony triangular crest which may have served to stabilize it in flight and assist in directing it, rather like an aircraft rudder. Several well preserved skeletons of this reptile have been obtained from the Cretaceous rocks of Kansas, U.S.

Opposite The model shown here is based on original eggs and skeletal material obtained from an expedition to the Gobi Desert, Mongolia. It shows young *Protoceratops* emerging from their shells under what would appear to be ideal conditions. In fact some sudden change must have occurred to prevent the complete hatching and the tiny skeletons were found alongside the remains of the eggs. *Protoceratops* lived during the very late Jurassic and early Cretaceous times in Asia and was the first of the horned, Ceratopsian dinosaurs, eventually giving rise to the *Triceratops* which developed in the Upper Cretaceous of northern states of America. *Protoceratops* was by no means as large as the American species, adults barely reaching 2 m. (6½ ft) in total length.

Right Struthiomimus is so called because it resembles the ostrich in general features. It had a small bird-like head on a long narrow neck. The elongate body was balanced on two slender but muscular hind legs and it carried its forelimbs in a drooped fashion in front of it. These were quite small legs with long toes which could be used to scoop up eggs from the crude nests of other reptiles. One form, closely related to *Struthiomimus*, called *Oviraptor* actually gets its name from its habit of thieving eggs. These reptiles were not particularly fierce and relied mainly upon their ability to run quickly from their predators. They lived during Upper Cretaceous times in North America and parts of Mongolia, preferring the sandy plains and humid lowlands.

Following pages Shortly after the rise of the *Iguanodon* in Europe, a similar herbivorous dinosaur called *Anatosaurus* appeared in North America. It was somewhat more specialized than *Iguanodon* and was certainly of a more curious shape. Its jaws were produced forward in an ugly overlapping fashion to form a sort of duck-bill. It is thought that it fed on large quantities of waterweed from the shores of inland lakes and probably cropped at the plantations with its shovel-like bill or beak. Skeletons in a good state of preservation have been found in the Upper Cretaceous deposits of the western states, U.S. On average, the reptile would appear to have been just slightly less than 4 m. (13 ft) in height.

8 The Evolution of Mammals and Man

The Cenozoic Era

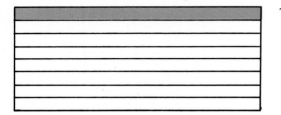

Geological history has been punctuated by outbursts of volcanic activity since the first conception of the planet within the universe. These outbursts have manifested themselves from time to time in various areas around the world. The effects of these active spots may radiate for many hundreds of miles around the original site, bringing changes in land and sea levels. The Cenozoic Era had many of these outbursts and, as a result, the conditions under which many of the animal populations were living changed considerably.

Some changes in land levels had already begun in Europe and North America at the end of the Mesozoic Era in the very late Cretaceous period. With them came a withdrawal of many sea areas and a complete change in deposition, with marked differences in the character of the sediments. The outstandingly pure white limestone of the Chalk seas was being overlaid in many cases by thin silty clays of vast river basins. Consequently, the once distinctly marine environment now became a more brackish or freshwater area, and the animals either had to adapt themselves to these new conditions or die out.

In time the continental uplift gave an added incentive to the struggling terrestrial life. Flowering plants, which made an appearance at the end of the Cretaceous, were now beginning to flourish, adding new dimensions to the scope of the fast-developing mammal populations.

In addition, the general world climate was once again changing. In the northern hemisphere a steady decline in temperatures began in the early Eocene and continued until the Pleistocene. This resulted in vast belts of both coniferous and deciduous trees developing, providing ample cover and food for the progress of numerous herbivores.

Thus with the Eocene, or the dawn of recent time, we begin to trace the development of the most adaptable group of animals ever known – mammals. It begins in western Europe and North America and reaches its zenith on the plains of South America and the Australasian continent.

Mammals can be roughly classified into two main groups. One of these, perhaps more familiar to us, is the group we call *placental*. These develop the foetal or embryonic stages of the young attached to a placenta or food reservoir within their bodies, eventually giving birth to a mature foetus or baby. A second group is called *marsupial* and most of the development of the immature young takes place in a specially provided pouch or pocket outside the mother's body. The young foetus is born within a very short time of conception and is transported, with considerable aid from its parent, to a pouch in which are the mammalian nipples from which it can suckle.

Throughout most of the early part of the Cenozoic both placental and marsupial mammals were fairly widespread, and fossil remains have been found in both Europe and the Americas. About the middle of the Cenozoic Era, the marsupials disappeared from the northern hemisphere and were confined to places in South America and the Australasian continent where they became isolated. During a long history of development they appear to have paralleled a great many of the adaptive biological niches of their placental counterparts. Their range of morphological species includes marsupial moles, mice, wolves, and even a sabre-tooth. They include both herbivores and carnivores among their numerous varieties.

Many of the earliest placental mammals were herbivores, the carnivores being represented in the Eocene by small specialized species, such as the creodonts. Two of the best examples of these primitive carnivores are *Hyaenodon*, which had teeth specially adapted for eating fresh meat, and *Pterodon*, which had teeth developed for eating carrion or meat which was not freshly killed.

The herbivores included the earliest known horse, *Hyracotherium*, often referred to as *Eohippus* or the Dawn Horse. This was a lightly built, four-toed animal with few of the features recognized in modern horse, save for the rather elongate head, wide nostrils, and short tufted mane on its neck. It frequented the shaded woodlands and mossy glades of the forests from western Europe to Colorado and Wyoming in the western states of U.S. Records of its size vary, but most well preserved skeletons average about 20 to 22 inches (51–56 cm.) in length and approximately 2 feet (61 cm.) in height. We can follow the evolution of the horse quite simply by studying the toes of the skeletons found throughout the Cenozoic. Although, starting off with four toes on each foot, it later developed only three, which was probably as a result of a change in environment to a firmer land area or to the plains where it had to run faster in order to escape its predators. Later still, with an increasing need for more speed, the hoof evolved from an over-development of the middle toe, the two remaining toes becoming smaller and of little use in locomotion.

Other herbivores developed throughout the early Cenozoic Era, some showing tendencies towards elongated snouts and tusk-like teeth. These were representatives of the Proboscidea or early elephants. As a rule they were clumsy thick-set animals with large heads, but they were not always big bodied. The oldest ancestor of this group is *Moeritherium*, which lived in Eocene times in Egypt and was no larger than a domestic pig.

It was not until very much later, in Miocene times, that the Proboscidea began to assume their characteristically enormous proportions. *Deinotherium*, one of the better known ancestors of the modern elephant, was a large bodied animal with big ears, a well developed trunk, and short, downward curving tusks on the lower jaw. It lived in Asia, Africa and it is probable that it was also found in parts of eastern Europe.

One species belonging to this group, the mighty mammoth, *Mammuthus primigenius*, has endeared itself to the civilized world. Its presence in any illustration creates in us a feeling of freindly recognition. In fact, this giant was a widely distributed representative of its group and many examples have been found in very good condition in the Pleistocene Ice Age deposits of Europe, Asia and America.

Early rhinoceroses are known to have existed in the Oligocene period, but these were without the characteristic nasal horn and were smooth-skinned. By the Pleistocene, they too had evolved and were also members of the Great Ice Age scene from Asia and Europe. The present-day rhinoceroses show very little change from their ancestral forms.

While Europe and parts of the North American continent were subjected to glacial action from the extension of the northern ice cap during the Pleistocene, the mammals of South America were developing in a very different way. *Megatherium*, the giant ground-sloth, was a huge bear-like herbivore approximately 20 feet (6·1 m.) long which had long claws on its flattened inturned feet and a

shaggy mass of coarse fur. Lumbering its way across the plains, it fed upon roots and leafy plants bordering the pampas.

Sharing the open plains of the South American continent was the armoured *Glyptodon*, which reached a length of approximately 12 feet (3·7 m.), and had a tough bony shell or carapace, short legs, heavily protected head and armoured tail. It was a burrower and used its long claws for digging. The present-day armadillo is probably related to this giant.

By the middle of the Quaternary times many genera and species of mammals had evolved. Some became highly specialized and grew large horns or antlers for defence. *Megaceros*, the great Irish deer, was a prime example of this development, having to carry the burden of an enormous pair of antlers almost 12 feet (3·7 m.) in total span. Many fine specimens of this enormous deer have been recovered from the Irish peat bogs where they have been since Pleistocene times.

Perhaps the most fascinating part of the evolutionary history of the mammals is the one which deals with the appearance of Man himself. The vanity expressed by *Homo sapiens* about his own beginnings is fundamentally a scientific enquiry, for most of us are anxious to know how it all began. The answer to this problem is still not clear and no really acceptable explanation for the origins of Man has so far been presented. There are, however, many ideas, most of them founded on logical scientific data.

Nowadays it is generally accepted that the apes and Man had a common ancestor within the Primates which gave rise to two, if not more, branches in the line of evolution. One of these branches became more specialized in certain anatomical details and is known as the Hominoid. It is to this group of rather specialized monkeys that Man, chimpanzees, and various other apes belong.

The special features which distinguish Man from the rest of the Primates are, in the main, confined to an increased brain capacity. The body of this unique animal has changed very little throughout its evolutionary history, although the assumption of an upright posture has lead to a shifting of internal weight and a difference in the structure of pelvic bones and limbs. This posture must have contributed very largely to the survival of some of the earlier species, since Man could never have been able to outrun his predators, but would certainly have had a clearer and greater vision from an upright stance.

Assuming, therefore, that Man is little more than a rather specialized ape, the question now arises as to when and in what form he broke away from the main stream. There is evidence that, during the Miocene period, there was considerable speciation among the apes which adapted themselves to various ecological niches. In all probability the hominoid line is to be found in one of these, but so far we have very little evidence of this development.

Recently, some anthropologists have attached a great deal of importance to a find, made at Lake Rudolf in Kenya, of a skull at present known by its registered number of 1470. It leads them to believe, once again, that there may have been two distinct and separate lines of development arising from a common ancestor. On the one hand, a more specialized strain represented by *Australopithecus*, found near Taung, South Africa, and on the other hand a contemporary line following from 1470 and leading to *Homo erectus* and on to *Homo sapiens* or modern Man.

The evolution of the Primates is almost a science in its own right and has been the subject of many scientific papers and books and cannot be dealt with or dismissed here. It seems inevitable that, with the coming of Man, the whole concept of prehistoric life must change and we find ourselves stepping from the role of unseen observer of the romantic and awe-inspiring history of animal evolution and immediately rising to the very pinnacle of that evolutionary tree. Masters we may be, but through our mastery and progress into a civilized and polluted world, we may yet reverse the tables of evolution and reduce the animal population to extinction.

Previous page The familiar sight of the friendly horse is an accepted part of our everyday life. The dignity and courage which this noble animal has displayed, from the thoroughbred to the muscular power of the working horse, has been an inspiration to us for many years. We might not have felt so inspired had the modern horse, *Equus*, remained like his earliest ancestor *Hyracotherium*, shown here in a diorama. This neat little mammal was no larger than an Alsatian (German Shepherd) dog and had soft-padded, four-toed feet adapted to life in the soft underglades of early Cenozoic forests.

Opposite From well-preserved fossil fish specimens, it has been possible to reconstruct models such as the catfish *Arius kitsoni* shown here. As we look at this model we may notice some aspects of its appearance which are not seen on the modern equivalent of the genus. They are in fact very closely related to the catfish that we find in our ponds and streams.

Below Pristigenys substriatus was a neat little bony fish that lived among the coral reefs in the Mediterranean area during the Eocene period. It is typical of the sort of tropical fish which live in the shallow warm seas of the Indian Ocean and Pacific. In this way, it is a very useful fossil because it tells us something about the climatic conditions which existed during the Eocene of that area.

Bottom When we look at the model reconstruction of this *Pristigenys substriatus* from the Eocene of north Italy shown here, we can see how very little has been lost in the preservation of the little fossil fish from Monte Bolca. The colour, of course, has been chosen by the artist who made the model, but it is probably very near the natural coloration which the species would have had while alive.

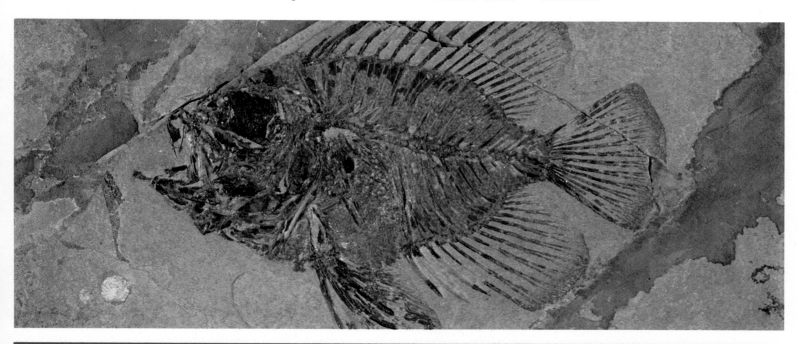

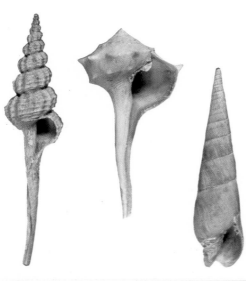

Below The rate of evolution of the land vertebrates appeared to be very much faster than that of aquatic animals. Many genera of fish which arose during the Jurassic period were still in existence in Cenozoic seas. Likewise, many of the bony fish which began to thrive in the Eocene times can be recognized in modern seas. They are represented by quite new species but are evidently closely related forms. The diorama represents life in an Eocene basin. It shows the predominant cartilaginous fish, the sharks and rays, with the background of smaller bony fish. In the foreground, however, is a large herring-like bony fish which was also fairly common at this time. Meanwhile, isolated groups of fish were adapting

themselves to more specialized conditions, such as those of the coral seas.

Left Towards the middle of Cenozoic times, the gastropods began to assume shapes and patterns which they still maintain today. Many of the genera which originated as early as the Eocene period are in evidence in modern seas and in many cases occur in relatively the same geographical areas as their fossil counterparts. The group pictured here are typical members of their respective genera. The large specimen in the centre of the group is a *Tudicula rusticula* and the specimen on the right is a *Terebra* (*Subula*) *plicaria*, both are from the Miocene of France. The rather ornate specimen on the

left is a *Fusinus porrectus* from the Eocene of Hampshire, England. All are approximately 7·5 cm. (3 in.) long.

Right In the Indian Ocean and also in parts of the Pacific, a giant clam has often become the subject of strange legends and stories. The clam referred to is the enormous *Tridacna* which sometimes reaches well over 1 m. (3¼ ft) in width. The specimen figured here, although closely related to the living species from the eastern seas, is in fact only 8·5 cm. (3¼ in.) in width and is a *Tridacna crocea* from the Pleistocene of Egypt.

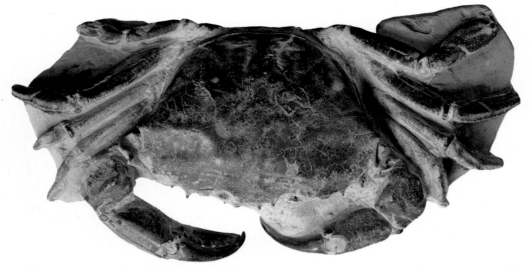

Top left Among the larger decapods, the crabs figure prominently. There are numerous species, some with fine spidery legs and delicate bodies, while others, such as this *Archaeogeryon peruvianus* from the Miocene of Patagonia, looks tough enough to stand up to quite a lot of punishment. This specimen is 18 cm. (7 in.) wide.

Centre left The thin-bedded limestones of the hills around Monte Bolca in Italy have been worked for many years to produce tiles and paving slabs. One of the by-products which has resulted from this industry is a magnificent collection of fossil fish. They all originate from the Cenozoic Era and are mainly from the Eocene beds. The specimens are so well preserved that they have become collectors' pieces and are now much more valuable than the limestone. *Eobothus minimus* was a neat little bony fish from the Eocene limestone of Monte Bolca, Italy. It was the earliest flatfish and is very closely related to modern flounders. The central spinal column supported numerous fine spiny ribs in an oval shape. There was a fringe of delicate dorsal fin and a similar ventral fin. The skeleton is about 6 cm. ($2\frac{3}{8}$ in.) long.

Bottom left Here is a specimen of *Naseus*, from Monte Bolca, an ornate tropical fish which lived among the coral reefs, probably feeding on the polyps in very much the same way as related species do today in the coral seas of the Indian Ocean and Pacific. The specimen is 14 cm. ($5\frac{1}{2}$ in.) long.

Below Holocentrus rubrum is a modern fish which lives in coral seas in the tropics. It is included here for comparison with the fossil *Naseus* beside it from the Monte Bolca limestones of Italy. *Naseus* must have looked very much like *Holocentrus* when alive and was certainly adapted to feed on the polyps of the corals from the reefs in very much the same way as the living species. The mouth

was small with slightly protruding lower jaw which enabled it to get near enough to the coral to nibble the animal inside.

Right Towards the end of the Cenozoic Era, many species which had evolved on the plains of South America either died out or found their way into North America by means of the narrow land strip between the two continents. One of these species was closely related to *Glyptodon* and has given rise to the armadillo which has spread through Mexico and Texas and some of the southern states. The charming little furry Armadillo show here is a survivor of this migration and is another example of a 'living fossil'. Although related to *Glyptodon*, the armadillo has a flexible carapace and can curl itself into a ball in defence.

Below During the late Pleistocene period the mammals had established a distinct pattern of global distribution. On the open plains of South America there were many species which were not to be found in any other parts of the world. One of these was the *Glyptodon*, an early ancestor of the armadillo which still exists in that region. It was considerably larger than the modern armadillo, reaching between 2 and 4 m. (6½–13 ft) in length, and had a hard fixed carapace or shell surrounding its body. Its head, which also had a small cap-like armoured shield upon it, had deep massive jaws and very distinctive teeth which it used for crushing.

Previous pages The sabre-tooth tiger was not a tiger at all but a large member of the cat family which must have looked rather like a puma. The species had a wide distribution throughout the North American continent during the Pleistocene, but we are more familiar with it as it is often depicted as the dominant predator at tar pit scenes such as Rancho La Brea in California. The extraordinary length of the canine teeth of this cat must surely have been used to stab its prey, although it did have an extremely wide articulation of its lower jaw. Most of the skeletons obtained stand about 1 m. (3¼ ft) in height. The correct name for this sabre-tooth is *Smilodon*.

Left The fine and delicate lace-wing flies seen here as museum specimens, might well have just settled on the limestone blocks in which they were found, they are so perfectly preserved. They are in fact fossil examples of a type of lace-wing, *Palaeochrysa stricta*, from Miocene deposits of Florissant, Colorado, U.S. The part and counterpart of the same specimen are 2 cm. (⅘ in.) long.

Below During the early Pleistocene period when a great deal of the northern hemisphere of the earth was experiencing the intense cold of the Ice Age, many groups of animals adapted themselves to these Arctic conditions. The Woolly Rhinoceros, *Coelodonta antiquitatis*, which is closely related to the modern species, became specialized in this

way by growing a thick shaggy coat of coarse hair to insulate it from the cold. It was a fairly widespread species, ranging from eastern Asia to northern Europe and Britain. Unlike the modern species, the Woolly Rhinoceros had a very long nasal horn and a well-developed secondary one and probably carried its head low. It was smaller than the modern species and reached an estimated maximum height of $1\frac{1}{2}$ m. ($4\frac{3}{4}$ ft) at the shoulder.

Below *Mammuthus primigenius* had a coarse shaggy coat to insulate it from the intense cold of the Quaternary Ice Age during which it lived. It was mainly to be found in the northern part of the European continent but ranged across Asia, making its way via

an Alaskan land bridge to the eastern flanks of the North American continent. There is plenty of evidence that early cave-man, living in the Stone Age, hunted the Woolly Mammoth for food. Many cave paintings depict this animal alongside other contemporaneous beasts.

Right The resin which oozed from the trees in Cenozoic times has hardened to amber over the years. Inside the amber we often find perfectly preserved specimens of animal and insect life which was alive during that period. Flies are particularly common victims of the sticky resin and entomologists have studied the specimens which have become trapped. They have found that there is very little difference between the species

which existed in those distant days and many of the species which are alive now. The specimen seen here is a *Rhagio* enclosed in a piece of Baltic amber. It is only 8 mm. ($\frac{5}{10}$ in.) long.

Previous pages Uintatherium was a large ungulate mammal living in the Eocene period. It had an elongated head upon which were six rounded horns. The teeth of this herbivore are of particular interest, since it had fairly long tusks projecting downwards from the upper jaw. It averaged about 4 m. (13 ft) in length and stood over 2 m. (6½ ft) high at the shoulder.

Right Megaceros is often referred to as the giant Irish Elk. It is not an elk but a species of deer and lived during the Pleistocene on the lush vegetation which grew just beyond the limits of the great ice sheets in the Ice Age. Skeletons, complete with the gigantic antlers, almost 4 m. (13 ft) in span, have been found in the peat bogs of Ireland. The species survived the Ice Age and probably lived on until early historic times.

Overleaf above The asphalt tar pits at Rancho La Brea in California have been the source of many fine, well-preserved skeletons of animals which lived in that area between 20,000 and 100,000 years ago. Obviously the exact way in which these animals became enclosed in the sticky mass will never be known, but we can imagine that grazing animals ranging from deer and ground-sloth to mastodon and mammoth were attracted to the water which sometimes collected on the surface of the tar and then became easy prey to the fierce predators, such as *Smilodon*. Sometimes victim and predator would become unfortunate casualties and both would sink. The remains of giant vultures with a wingspan of some 3 to 3·7 m. (10–12 ft) have also been found in the tar.

Overleaf below left If we look at the evolutionary sequence of some animal species, we may recognize certain features which immediately strike us as familiar. There can be very few of us who would not know an elephant when we saw one, but the process of evolution towards this animal has been long and varied and many branches or near species have not been successful. One of the evolutionary stepping stones on the way to modern elephant was *Deinotherium* which lived in eastern Europe and Asia in Miocene and Pliocene times. It was a large animal, well over 4 m. (13 ft) high and lived on the soft vegetation of small trees and bushes which it chewed with its simple cheek teeth. One very distinctive feature of this beast is the position of the short stout tusks which, unlike modern elephant, are curved downwards in a fairly low arc.

Overleaf right The amount of interest which surrounds any discovery which has a bearing on the evolution of Man is immense. When a skull with what appeared to be advanced hominoid characters was discovered in 1948, it was thought to be a very early ancestor of Man. This skull was described as *Proconsul africanus* (top) and was found in Miocene deposits in East Africa. Subsequent study of the skull suggested to many anthropologists that it was no more than a chimpanzee-like ape which possibly spent some of its time in an erect posture and may not have been in any way ancestral to Man. While doubts

have been expressed about the ancestry and evolutionary trends of many primitive hominoids, there can be little doubt that both Peking Man and Java Man (*centre*) from the Lower Pleistocene period were of a comparatively advanced species. They emerge, therefore, as true Man rather than as man-apes. It is still not quite clear as to how these eastern species relate to the African Australopithecines. Both Peking Man and Java Man should be regarded as belonging to the same genus, that is *Pithecanthropus*. This was a heavily browed, square-jawed man, probably not more than 1·5 m. (5 ft) high. The long journey from the dawn of the mammals to modern Man has had many turnings. Some of these have led to the development of numerous species of both carnivore and herbivore but the main stream which eventually led to the evolution of Man must have been through the apes. With the additional powers of reasoning and the ever-increasing brain capacity it followed that the very shape of the skull must change and, with it, the features of the individual. One of the earliest hominoid apes which we can recognize as Man occurred during the Pleistocene period in Africa. This ancestor of ours was one of the first specialized apes to walk upright constantly and to use roughly fashioned tools and weapons. We call him *Australopithecus* or the southern ape-man (*bottom*). From information obtained from examination of the skulls of this being, we are able to reconstruct the sort of image that this primitive individual must have presented. He was certainly more ape-like than any subsequent human forms and had well developed brow-ridges and a low, square, projecting jaw. The specimen is a cast of *Australopithecus africanus* from South Africa and is without the lower jaw.

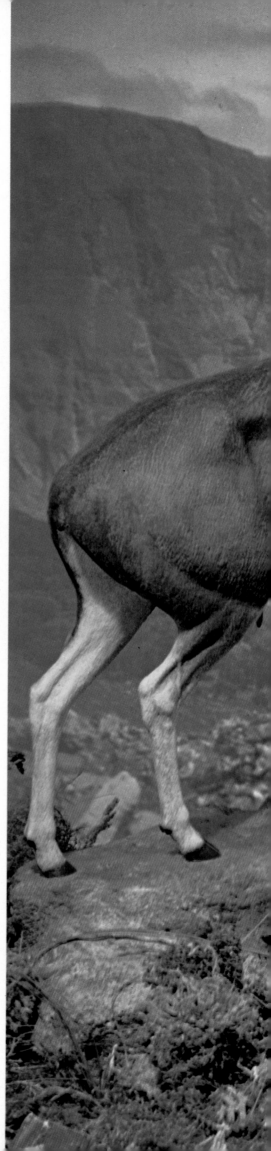

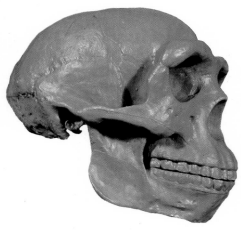

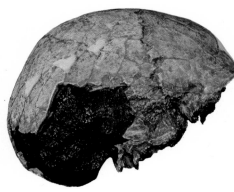

Index

Acknowledgments

The publishers would like to thank the following organizations and individuals for their kind permission to reproduce the pictures in this book:

Heather Angel: 7 above, 13 below, 14 below, 67 below left
Ardea: (Arthur Hayward) 59 below, 66, 74–75
British Museum (Natural History) (Photo: Imitor): jacket flaps, 10–11, 13 above right, 14 above, 14 centre, 15 above left, 15 above right, 15 centre left, 19, 21 above right, 21 below right, 22 above, 23 above left, 23 above right, 23 centre, 24–25, 26 centre, 26 below, 27 above, 30 above, 31 centre, 32, 32–33 below, 33 above, 35, 36, 37 above, 37 below, 40–41, 43,
44 above left, 45 right, 46 below, 52, 53 centre, 53 below, 54–55, 56 above left, 56 above centre, 56 above right, 56 below, 57 above, 57 below, 58 above, 67 above, 67 centre left, 70–71 below, 71 above left, 71 above centre, 71 above right, 71 below right, 72 above, 72 centre, 73, 79, 81 above, 82 above, 82–83 below, 83 above, 84 above left, 84 centre left, 84 below left, 85 below, 88 above, 89 above, 94 above, 94 below, 95 all; (Photoresources): 67 below right, 80, 81 below.
European Colour Library: 4–5, 11 above, 13 above left, 16 right, 21 left, 26 above, 38–39, 53 above, 58 centre.
Horniman Museum, London (Photoresources): jacket front, 62–63. N.H.P.A. (H. R. Allen): 2–3

(Rhamphorhynchus, Royal Scottish Museum), 44–45 below (Royal Scottish Museum).
Natural Science Photos: 14–15 below (F. Greenway), 22–23 below, 31 above, 44–45 above; (Arthur Hayward) jacket back, 9, 12, 16 left, 20, 22 below, 27 below, 29, 30 below, 48–49, 51, 60–61, 65, 68–69, 72 below left, 76–77, 88 below, 90–91, end papers, (G. Kinns) 10 above, 59 above, 86–87, 88 below, 89 below, 90–91, 92–93 (P.Ward) 17, 46 above.
Photoaquatics: 84 below right
Z.E.F.A.: (Hemlinger) 47 (Luden) 58 below.
Zoological Society of London: 85 above.